MODELING THE TRANSITION ERA

Tony Koester

WAUKESHA, WI

ACKNOWLEDGEMENTS

It takes a sizable support group to produce a book on a multifaceted topic like the steam-to-diesel transition era. I extend my sincere appreciation to those who provided information and photos, including Gerry Albers, Tom Bailey, Rich Bourgerie, Mike Brock, Ken Buck, Bruce Carpenter, Chesapeake & Ohio Historical Society, Warner Clark, Dave Clemens, Gregg Condon, Edwin Cooper, Bill Darnaby, Rick De Candido, George Dutka, Bruce Ernatt, Chuck Geletzke, Gene Green, Doug Gurin, Doug Harding, Dan Holbrook, Dale Jenkins, Barry Karlberg, Bert Kram, Jim Kreider, Randy Laframboise, Doug Leffler, Hal Lewis, Gordon Locke, Bruce Meyer, Charlotte Schwab Miller, Bob Mohowski, Seth Neumann, Jeff Otto, Jack Ozanich, Ted Pamperin, Karen Parker, Cliff Powers, Clark Propst, Jim Providenza, Bill Raia, John Rogers, Andy Rubbo, Mike Schafer, Bill Schaumburg, Karl Schoettlin, Jim Semon, Sam Sherman, Wayne Sittner, Jim Six, Mont Switzer, Doug Tagsold, Tony Thompson, Harold Werthwein, Allen Whittaker, Aubrey Wiley, Craig Wilson, and Ted York. Special thanks are extended to Jeff Wilson, Diane Bacha, and Randy Rehberg, who guided this book through Kalmbach's meticulous production process.

Tony Koester

ON THE COVER

Electro-Motive's game-changing 1,350-hp FT locomotive helped win World War II and established the beachhead that vanquished steam.
Photo by Rick De Candido

Photographs by the author except where noted.

Kalmbach Books
21027 Crossroads Circle
Waukesha, Wisconsin 53186
www.Kalmbach.com/Books

Published in 2017
21 20 19 18 17 1 2 3 4 5

Manufactured in China

ISBN: 978-1-62700-493-0
EISBN: 978-1-62700-494-7

Editor: Jeff Wilson
Book Design: Lil Weber

Library of Congress Control Number: 2017941410

CONTENTS

The Electro-Motive FT (originally the "Model F") diesel-electric locomotive, right, was subdivided into 1,350-hp A units with control cabs or B units without cabs. Many A-B and even some A-B-B-A sets were semi-permanently coupled with drawbars rather than couplers, thus making a single 2,700-hp or 5,400-hp locomotive that could be operated by one engineer and "fireman" without violating early labor agreements concerning one crew per locomotive. Agreements that diesel locomotives operated in multiple-unit sets constituted one locomotive became the norm. *Rick De Candido*

INTRODUCTION

A unique niche in history

"It was September 1954 on the Milwaukee Road's Shullsburg Branch. My uncle John said there would be a transition era in which both steam and diesel would rule the line. But it was a transition midnight: The three Ten-Wheelers out on the line and its branches all went to Janesville, Wis., one fall day as seven-year-old me watched. Next day three diesels came, and that was that."—Gregg Condon

Before the Electro-Motive Corporation (EMC) was merged into General Motors (becoming GM's Electro-Motive Division, or EMD), it sent a four-unit FT demonstrator set to be tested on the New York Central in September 1940. Rick De Candido paid tribute to that event in HO using Stewart models he painted and decaled. This scene was reenacted numerous times across the land as sleek diesel-electric cab and booster units and later boxy hood units sidled up to the snorting, hissing beasts they would soon displace. *Rick De Candido*

A one-time opportunity

We'll never see such a time again. A type of railroading and the massive physical plant built to support it vanished in two decades. The new kid on the block, the diesel-electric locomotive, transitioned from a novelty to an unstoppable tide of mass-produced components and entire assemblies. The quaint term "unit" became the norm, and for good reason: It was the key to the diesel's flexibility, its ability to do any job, any time, anywhere. And all too soon for the steam fan, it did so with remarkable economy, reliability, and stamina.

What was a revolution for the railroad industry was an unmitigated disaster for the rail enthusiast. Even to the most disinterested observer, a steam locomotive is a living, breathing machine, panting and snorting, hissing and dripping. The vanquisher certainly has its attributes, but it is clearly a rather simple machine—itself an attribute—although some examples came close to evoking the visual and aural drama of steam. Recall the late George Hilton's description in *Trains* magazine of a chugging, smoke-emitting Alco PA-1 passenger locomotive as "an honorary steam locomotive." Indeed!

Younger modelers who missed the end of the steam locomotive's reign often fail to understand or appreciate the overwhelming impression steam could impart. One locomotive pulling a motley assemblage of passenger cars on a fan trip is a delight for sore eyes but doesn't come close to what one experienced when encountering a nest of these beasts in an engine terminal, or a massive 4-8-4 at speed overtaking a diminutive-by-comparison 2-8-2 tucked safely out of harm's way in a passing track. The dynamics were incredible and never again to be repeated.

Or will they? In fact, we modelers do an amazingly good job of building time machines. I can walk from my home office only a few yards to the basement stairs and then descend into a time capsule that allows me to step back into the west-central Indiana and east-central Illinois of 1954. There, patiently waiting for me to turn on the lights and power, are trains that are frozen in time from last month's "operating session" when they were busily re-creating a golden year of the transition era, which I'll loosely define as from the end of World War II to 1960, although the debut of Electro-Motive's FT in late 1939 kick-started the paradigm shift. Shortly, they will be back at work, and—thanks to Digital Command Control and myriad manufacturers of very finely molded plastic, cast metal, and shaped brass models and sound decoders—hissing and chuffing and whining and doing all of the other things (except, with any luck at all, issuing black smoke) that their prototypes once did.

The transition era remains by far the most popular time to depict on our model railroads. That's understandable, as it was railroading's most colorful and variegated period. But that's also remarkable, as that era's fires died more than a half-century ago, long before many of those who model it were born!

What makes this time so remarkable and model-worthy is what we'll explore in this book. The era of big steam, colorful first-generation diesels, the debut of "covered wagons" and "hood units," the time when manned depots and interlocking towers dotted the landscape and cabooses marked the end of every freight train hasn't disappeared at all. It's just been scaled down.

A 3800-class 2-10-2 shoves a Green Fruit Extra up Cajon Pass as a set of F units eases the Santa Fe's *Super Chief* down the roughly 2 percent grade in this HO scene. The whine of the diesels' dynamic-brake fans is all but drowned out by the powerful chuffs of the steam locomotive struggling upgrade. *Ted York*

CHAPTER ONE

Defining the Transition Era

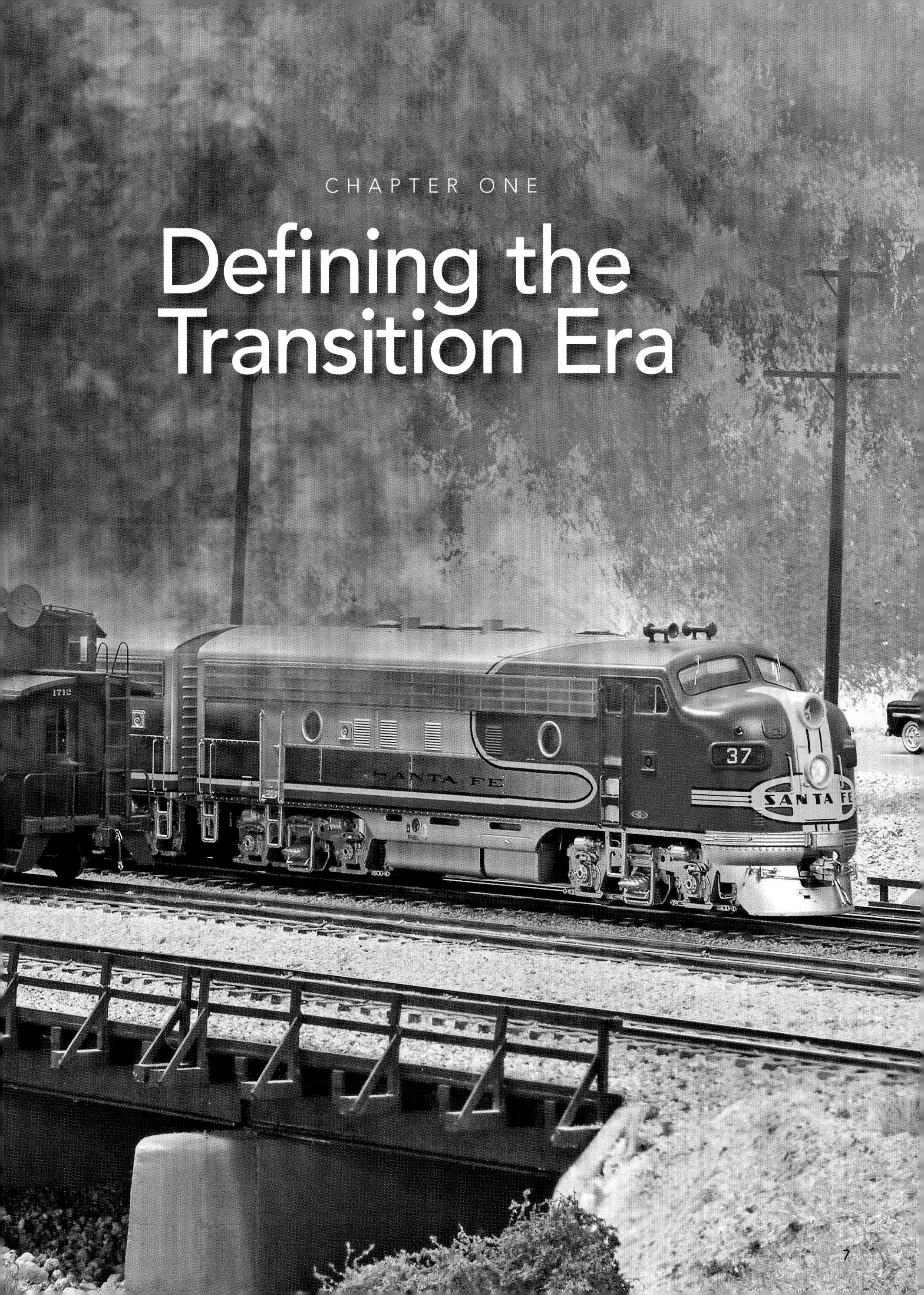

The "Transition Era," as the U.S. and Canada segued from the global and then Korean conflicts to a peacetime economy, continues to be the most popular time span to model. There are model-worthy periods on either side of that pivotal period, but none of them offers the same broad base of characters. For a while, steam and diesel and even electrified railroading coexisted relatively peacefully. Illusory or not, this seminal period affords myriad modeling opportunities.

Grasping the situation

It's difficult to understand an era without some grasp of what preceded and followed it. Moreover, it's simplistic to consider the entire period from just after World War II to 1960 as one era. Railroading in the late 1940s was in many instances quite different from railroading in the late 1950s.

But this period shared one overriding characteristic: Both steam and diesel locomotives performed the tasks of everyday railroading, and both did them well, **1**. That's why this period remains so popular with modelers, a have-your-cake-and-eat-it-too situation.

Note, too, that this is not a history book. In the 100-odd pages we have to spend together, there's not much space to devote to topics that are only peripherally related to railroading. And you won't find a chapter devoted to modeling Transition Era freight cars. That's a topic so vast that it requires a book of its own, which Jeff Wilson has fortuitously provided: *Freight Cars of the '40s and '50s* (Kalmbach, 2015).

So let's content ourselves with a brief look back beyond the post-war steam-to-diesel Transition Era and perhaps an occasional glimpse beyond steam's waning hours in 1960.

The 1949 and '50 Fords had a single "bullet" in the grille (left), whereas the 1951 models (right) had a pair of them. *Jim Semon*

Signs of the times

The Depression was a bad memory blotted out by an even worse one: World War II. Everything and everyone was tired and worn out. Olive drab was still the predominant color, if only in deeply etched memories. Prior to 1948, "new" vehicles were in fact built to pre-war specs hidden under brighter colors.

But as the dreary war years began to fade from public consciousness, everyone was eager for something new and bright. In 1949, Ford eschewed bulging separate fenders with the classic 1949-51 "Shoebox," **2**. Then came Rock & Roll, two-tone automobiles, **3** (my dad bought a stunning orange-and-white 1956 Ford hardtop), blue suede shoes, 45-rpm records, and fuzzy black-and-white televisions.

My good modeling friend John Rogers, a psychiatrist, puts it this way: "It might be useful to think about the Transition Era as a cultural phenomenon. It was the heyday of the railroads because the technology was bringing better rail transportation and we were not yet fully into the airplane era. I think the flashy diesels and swank cars are a reflection of the era they served."

Despite Cold War concerns, the good times won out as we discovered that the biggest tail fins in town were not on enemy bombers but on Plymouths and Dodges and Cadillacs and, most notably, the 1957 Chevrolet, **4**. The big news in late 1954 was that Chevy finally had a V8 engine and, borrowed from the pricier GM cars of a year or two before, a wrap-around windshield, **5**. Ford launched the Thunderbird, countered by the fiberglass-bodied Corvette. "Dream cars" from all the major builders showed us that Tomorrow would be even better.

By the 1950s, trains like the Burlington's *Zephyrs* that had ushered in the streamliner era two decades before, **6**, were barely hanging on. Even GM's bus-body-based Aerotrain, **7**, was soon relegated to commuter service in Chicago. Its very light weight translated to a rougher ride, and adding more cars to a train consist was no easy task. New fleets of lightweight streamlined cars from Pullman and Budd replaced war-weary heavyweight consists.

From a pro's perspective

Chuck Geletzke, an active modeler as well as a recently retired railroader, just missed the Transition Era and end of steam by seven years when he hired out on the Grand Trunk Western in 1967. As it turned out, he hadn't missed much after all:

"When I started my career, we still had Morse telegraph on the branch lines, car ferries, iced reefers and icing stations, kerosene lanterns on locomotives, and wood cabooses without toilet facilities. We learned the art of working with hand signals (we didn't start getting radios until 1968), using them while sometimes standing on top of the cars on running boards. We worked with switch lists and waybills.

"There were depots with agents and clerks, towers with operators and levermen, elevated towers or shanties with crossing watchmen. We rode in cabooses behind stinky stock cars. Passenger trains were owned and operated by the railroad featuring heavyweight cars, express reefers, diners, lounge cars, and Pullmans.

3 The 1955 and similar 1956 (shown here) Fords helped bring in the age of splashy two-tone paint jobs and sleek lines. The tailfin was just beginning to make an appearance.

4 The 1957 Chevy is a classic example of 1950s styling. Even the lowly station wagon—Nomad in Chevyspeak—had tailfins and a powerful small-block V8 engine.

5 The 1955 Chevrolet introduced an updated body style (but still retained a hint of separate rear fenders) with a wrap-around windshield and, at long last, a V8 engine to compete with Ford's new overhead-valve V8.

6

Boston & Maine's *Cheshire*, nee *Flying Yankee*—a copy of Burlington's successful *Zephyr* built by Budd—passes Webb, N.H., on its way from Boston to White River Junction, Vt. (above). On different routes, it acquired various names such as *Cheshire, Minuteman,* and *Business Man*. Rich Cobb built the depot. The noses of Burlington's E units (left) were painted to evoke memories of the railroad's famous *Pioneer Zephyr.* Note the painted "grilles" alongside the headlight and the horizontal bars representing the windshield panes. *B&M: Jim Dufour; CB&Q: Trains magazine collection*

"Even in the 1960s," Chuck continues, "you could tell railroads apart by looking at the right-of-way. There were both car and locomotive shops and roundhouses; cars were weighed on scales; four and five men comprised a crew; yard jobs were eight hours; road jobs were paid by the mile and 100 miles was the 'basic day'; the 16-hour day was still the norm; all of our rail was jointed; occupied outfit or camp cars were still used, as were semaphore block and train-order signals; there were lineside dispatcher phones; we could kick and drop cars; there was no reflectorized clothing; railroad approved watches were required; employee timetables and rule books would fit in your pocket; passenger trains handled U.S. Mail; and the local moved LCL (less-than-carload lot) freight."

Small industries

One of the most important characteristics of the Transition Era was that myriad small "industries"—spots where freight cars can be loaded or unloaded—were located in every town. These included lumberyards, coal yards, petroleum bulk dealers, team tracks, freight houses, grain elevators, brickyards, and especially interchanges with other railroads.

Interchanges are important in that they are "universal industries": Every type and quantity of freight car may show up on an interchange track, making them ideal modeling candidates. (See my book *Space-saving Industries for Your Layout,* Kalmbach 2016.)

When combined with railroad structures such as the depot, freight house, interlocking tower, and perhaps a water tower, these physically and operationally attractive buildings are alone sufficient justification to model the transition era.

Woodard's revolution

That modern steam locomotives performed superbly into the 1950s is due in no small part to a mechanical engineer named Will Woodard. He worked for Lima Locomotive Works in Lima, Ohio (This Lima is pronounced like the bean, not "lee-ma," as in Peru.) Woodard came up with the idea of what came to be called "Super Power," wherein a steam locomotive could develop more steam than it could possibly use up. Simply put, rather than design a steam locomotive either for

7

General Motors' *Aerotrain* was an experiment in lightweight construction using the shells of GM motor coaches. Unpopular due to their rough ride, the trains wound up in Rock Island commuter service in Chicago. This one resides in the National Railroad Museum in Green Bay, Wis.; another is at the Museum of Transportation in St. Louis.

8

The era of truly modern steam power dates to Lima's development of Super Power and the Boston & Albany Berkshire (2-8-4), which hauled freight through its namesake mountains. But the B&A locomotive's small (63"-diameter) drivers failed to take advantage of its boiler to generate steam fast enough for high-speed service. By the transition era, such machines typically had 69" or larger drivers. Number 1421 is eastbound at Warren, Mass., on October 14, 1947. *Robert A. Buck, Kenneth J. Buck collection*

The first Nickel Plate Berkshires relied heavily on work done by Will Woodard at Lima Locomotive Works, but Alco won the contract for the first 15 (nos. 700-714). These locomotives, including 708 climbing out of the Wabash Valley in West Lafayette, Ind., served admirably from 1934 to the end of mainline steam in 1958. *Hal Lewis*

pulling power or for speed, Woodard's Super Power locomotives could move tonnage at speed.

The first example of Super Power was a series of A-1 class 2-8-4s built for the Boston & Albany in 1925, **8**. The B&A used them to move tonnage through the Berkshire Mountains in western Massachusetts, hence the name "Berkshire."

What started at Lima spread to American Locomotive Co. (Alco) and Baldwin. An early and important step forward taken in the quest for faster, more-powerful, and sustainable-at-speed steam power was the Lima-built Chesapeake & Ohio 2-10-4 (1930) and the Nickel Plate Road Berkshire (1934). The NKP machine was basically scaled-down copy of the C&O T-1. The NKP Berkshires had 69" rather than the B&A Berkshire's Mikado-size 63" drivers, critical to fully utilizing Super Power's potential. But Alco, **9**, not Woodard's employer Lima, won the contract! Lima got even by winning all subsequent NKP Berkshire contracts, and as a result was the last of the Big Three commercial builders to deliver a steam locomotive in the U.S., NKP S-3 779 (in 1949), now fittingly on display in a park in Lima.

Ironically, Baldwin's last steam locomotive, also in 1949, was Chesapeake & Ohio 2-6-6-2 1309, **10**, a design dating back to the teens. It was recently restored for tourist-train service on the Western Maryland Scenic Railroad (wmsr.com) between Cumberland and Frostburg, Md.

The beginning of the end

I've marked the start of the Transition Era as 1946, just after World War II. The 1939 debut of Electro-Motive Corporation's 1,350-hp FT diesel suggests a starting date just prior

C&O 2-6-6-2s nos. 1309 and 1302 are departing Scarlet, W.Va., in June 1950 with a loaded coal train. Number 1309 and nine identical H-6s were the last steam locomotives built by Baldwin Locomotive Works (1949). Number 1309 has been restored for service on the Western Maryland Scenic Railroad in Cumberland, Md. *Gene Huddleston; courtesy C&O Historical Society*

to WWII, but the transition didn't really become clear as the wave of the future until the war was over. And even then, steam held on tenaciously; witness NKP's 1949 order for ten more Berkshires from Lima, this after testing F units, and Norfolk & Western continuing to build steam into the 1950s. It wasn't until the mid '50s that any reasonable observer would be prompted to write steam's obituary.

The development of new steam locomotive designs was frozen during the war in favor of copying proven designs while diesel-electric innovation was pushing forward. That explains 2-10-4s on the fiercely independent Pennsylvania that were mechanical clones of C&O T-1 2-10-4s but with PRR-style cabs, front-end treatments, and tenders, **11**.

Diesel's tentative beachhead was enough. There was no dislodging internal combustion after the war. The unitized FT was a remarkable development. Need more tractive effort or horsepower? Add another unit, not another locomotive with a separate crew—a critical difference as many railroaders were summoned to the war effort, and after the war as labor rates climbed.

Railroad brotherhoods were not oblivious to the implications, and one cannot examine any aspect of prototype railroading without considering labor agreements. Unions exist to protect the interests of their members, and in the case of the railroad industry, they resulted from callous and dangerous practices instigated by the barons of the 19th century who owned the railroads.

One can argue that at times the scales tipped too heavily in either direction, but for our purposes we'll simply acknowledge that one cannot ignore labor agreements and develop even a rudimentary understanding of why things turned out the way they did in the rail industry.

And so it was with the FT. From an engineer's (and his union's) standpoint, it could be argued that each unit in the consist of a diesel-electric was a "locomotive." The easy way around this was to permanently couple the cab, or A, unit to a cabless B (booster)

11

Wartime restrictions on new locomotive designs forced the highly independent Pennsylvania Railroad to copy the C&O's 2-10-4s; even a Pennsy cab and massive tender couldn't disguise their lineage. This offers an example for freelancers: Change details on a commercially available model to achieve a family appearance. Broadway Ltd. offers HO models of both the C&O T-1 and PRR J1. *Trains magazine collection*

12

Meat trains like the Nickel Plate Road's hot KC-44 (which usually ran on No. 90's schedule) and MB-98 (No. 98) between Madison, near St. Louis, and Buffalo, N.Y., typically had strings of bright red Swift reefers hanging on for dear life behind a fleet-footed Berkshire's tender—as they do today on the author's HO railroad.

13

Andrews trucks, which allowed reuse of the journal boxes from arch-bar trucks, were banned from interchange in 1957 but still permitted on locomotive tenders. *Lima Locomotive Works*

unit using a drawbar rather than a pair of knuckle couplers. Look carefully at photos of FT consists, and you'll notice that some of them appear to be coupled very closely together. That pair of units was designated a locomotive, and even a common numbering system—four units numbered 101A, B, C, and D, for example—bore out this subterfuge. But common sense and practicality won out, and the principle that a consist of diesel-electric units running in multiple with one crew was a single locomotive became accepted.

The "hog law"

The recognition that humans have endurance limits and need rest between runs had become law as the Railroad Hours of Service Act in 1907. (It wasn't substantially revised until 1969.) The resulting 16-continuous-hours-of-service limit was dubbed the "hog law." (See Chapter 9 for details and modeling implications of the hours-of-service law.)

Manual labor

Today it's hard to relate to the era when manual labor was so cheap that buying a labor-saving machine was considered expensive. Consider, for example, the days when heating one's home was most often done with soft (bituminous) or hard (anthracite) coal. How did the coal get from mine to home furnace? I suspect you're forming a mental picture of endless streams of two- and three-bay hopper cars oozing out of the mountains and heading singly and in pairs to Anytown, U.S.A. Think again.

Most small-town and small-city coal yards had one or more coal bins with very distinctive architecture with unloading doors high up on the track sides (see Chapter 8). There wasn't room for an unloading pit and conveyor between the track and the coal shed. Instead, the local coal dealer would cut a deal with a couple of high school kids or day laborers to shovel the coal by hand from a gondola into the shed. The going rate was around 50 cents per day! Why buy a conveyor and build a concrete unloading pit that would sit idle most of the time anyway?

This philosophy extended the life of the common boxcar. A carload of lumber would be shoved into the lumberyard or onto a nearby team track and unloaded one "stick" at a time into a bay or onto a flatbed truck. Bulkhead or center-beam flatcars quickly unloaded by one person driving a forklift were then not even a figment of someone's fertile imagination.

Another example of the boxcar hanging on beyond its best-if-used-by-date was found at every grain elevator. A worker nailed several "grain doors"—typically three horizontal boards joined with vertical slats—up to the top of the "field" side of the car. He'd then nail more grain doors to the elevator side of the boxcar, leaving just enough open space at the top to slide in the loading spout—and to escape.

When the loaded car reached a mill, soybean plant, or storage facility, it was unloaded by the reverse process. Adding to the cost and inefficiency, the grain doors were stenciled with the reporting marks of the sending railroad—they wanted them returned.

Disposable cardboard grain doors eventually displaced the wood doors, and then covered hoppers replaced boxcars—but not until the 1960s. Although two-bay covered hoppers date back to the 1930s, they were used for cement and lime, not grain.

Capacities and clearances

A knowledgeable modeler can tell at a glance whether a model railroader

Nickel Plate Road St. Louis Division timeline

Event	1951	1952	1953	1954	1955	1956	1957	1958	1959	1960	1961	1962
Mars lights on 2-8-2s, 2-8-4s	>>>>	>>>>	>>>>	>>>>	>>>>							
Steam to Peoria	>>>>	>>>>	>>>									
Steam to St. Louis	>>>>	>>>>	>>>>	>>>>	>>							
GP7s			>>	>>>>	>>>>	>>>>	>>>>	>>>>	>>>>	>>>>	>>>>	>>>>
RS-3s				>>>	>>>>	>>>>	>>>>	>>>>	>>>>	>>>>	>>>>	>>>>
GP9s					>>	>>>>	>>>>	>>>>	>>>>	>>>>	>>>>	>>>>
RS-11s						>>>>	>>>>	>>>>	>>>>	>>>>	>>>>	>>>>
New NKP paint scheme									>>>>	>>>>	>>>>	>>>>
Train Nos. 9 and 10	>>>>	>>>>	>>>>	>>>>	>>>>	>>>>	>>>>	>>>>	>			
GP18s										>>	>>>>	>>>>
Bay-window cabooses												>>>>
RS-36s												>>>>
GP30s												>

14 This table was helpful when I was choosing a specific year to model. My key dates were the arrival of personal-favorite Alco RS-3s (April 1954) and the demise of steam (July 1955) on the NKP's St. Louis Division. My decision: Autumn 1954, which was the last year that steam powered the fall grain rush. The photo of Berkshires and RS-3s (under coaling tower) at Frankfort, Ind., shows that ideal interval. *Darrell Finney; Mike Finney collection*

aspiring to depict the transition era has done his or her homework as a train passes by: If the heights of the boxcars in that train vary considerably, that's a good sign someone has been paying attention.

Varying colors? Not so much. Some hue of red-oxide was a cheap and durable paint; farmers even used it on barns. Today, we think in terms of boxcar red, oxide red, or Tuscan red, even freight car brown. We argue endlessly about which flavor thereof decorated our pet railroad's boxcars in March 1953. A casual observer might not see the differences, especially when you apply a healthy coating of soot and grime left over from the low-maintenance war years.

Even the "yellow bellies"—the endless streams of orange Pacific Fruit Express and yellow Fruit Growers Express reefers filled with fruit and vegetables and yellow Armour meat reefers—added only a hint of color to the typical freight. Thank goodness for Swift's bright red wood and then steel fleet, **12**, and an occasional aberration like a Bangor & Aroostook red, white, and blue boxcar (but see **9-10**).

Several members of the Steam-era Freight Car group compiled a list of various rulings, decisions, "bans," and "recommendations." (To inquire about membership, go to STMFC@yahoogroups.com.) Among their findings: Truss rods were never actually banned, but all-wood underframes had a drop-dead date of late 1928. Railroads were given several years either to scrap their all-wood underframe cars, or to upgrade them with new steel center sills and coupler pockets.

"That's why there was a flurry of new car orders in the mid-1920s," freight car researcher Ray Breyer reports, "especially for large single-sheathed cars (which were cheaper than double-sheathed or all-steel). Since the '40-year rule' for freight cars wasn't in effect until 1970, older cars with truss rods continued in use, albeit in quickly diminishing numbers, until the early 1960s.

"Arch-bar trucks were barred from interchange (but not maintenance-of-way) service by July 1940," Ray reports. "But WWII messed up that plan, and a car with arch bars could occasionally be seen wandering the national system as late as 1945. Freight-car authority Richard Hendrickson sent me a picture of an ancient Santa Fe 36-foot boxcar in Florida in 1944."

Ray adds that K-type air brakes were barred from interchange service, with a six-month grace period, between July 1953 and January 1, 1954. "After September 1933, no new cars were to be built with K brakes, and by January 1937 no rebuilt cars were to have Ks."

In 1957, freight car trucks with T-, L-, and I-shaped cross-sections were banned because they were prone to cracking. T-section Bettendorf and early Andrews trucks are included with some model freight cars, so this cutoff date has modeling implications. Freight-car historian Tony Thompson notes that the ban did not apply to Andrews U-section trucks such as those made in HO by Accurail, Bowser, InterMountain, Tahoe, and Walthers.

That year also saw the ban of truck sideframes that had bolted-on journal boxes, which included the popular Andrews type. The ruling didn't apply to cars that stayed online, including tender trucks, **13**, as they were not used in interchange service.

Richard Hendrickson prepared a superb overview of truck types offered by model railroad manufacturers: docs.google.com/file/d/0Bz_ctrHrDz4wcjJWcENpaDJYbUU/edit.

A year later, cast-iron wheels were prohibited on new or newly rebuilt cars. Most wheels with the curved ribs on the back were cast iron. And in 1959, a spate of derailments caused the banning of the popular Allied Full-Cushion high-speed truck.

The streamliner era

The Chicago, Burlington & Quincy's lightweight, fast, shovel-nose *Zephyrs* provided the first clear picture of the era of stainless-steel passenger trains. Indeed, one could see the "face" of the original Budd-built Zephyrs painted on the nose of the Q's EMD E units, **6**. The *Zephyr* design was copied by the Boston & Maine on its *Flying Yankee.*

Alco PAs. Thanks in part to Lionel's popular train sets, Santa Fe's warbonnet livery, **1**, became America's most recognized diesel paint scheme.

In the East, the flamboyance epitomized by Santa Fe's red noses, the Southern Pacific's *Daylights,* and the equally bright liveries applied to cab units of the Deep South, **7-1**, failed to gain a foothold. The New York Central and Pennsylvania maintained a conservative image reminiscent of pinstriped business suits and somber Tuscan in keeping with what the business crowd expected for trains leaving Manhattan.

Before long, however, it became clear that businessmen weren't riding the trains. The CZ was a land version of a cruise ship and attracted more tourists than business clientele. The *Super Chief* had hosted movie stars; now they preferred to travel on TWA Super Constellations and United DC-7s. By the mid-1950s, Boeing 707s and Douglas DC-8 jetliners were selling like hotcakes.

Despite heroic efforts to maintain a truly first-class traveling experience by Santa Fe, for one, the lightweight streamliner was increasingly on life support. And as mail contracts moved to trucks, even all-coach trains no longer made economic sense.

15

In 1957 the New York, Ontario & Western, faced with declining coal and bridge traffic, was the first major railroad to be abandoned. *Jim Shaughnessy*

The end of World War II brought with it a renewed optimism in the future and a need to replace just about everything worn out during the war. Surely passengers would continue to flock to the railroad station, just as they had when gas and tires were rationed and cars were aging pre-war models. This led railroads to order not just new passenger cars but entire trains. The Rio Grande, along with the Burlington and Western Pacific, debuted the Vista Dome to give passengers the same view enjoyed by the engineer and fireman as they traversed the Rockies and Sierras.

Like the *California Zephyr,* Santa Fe's *Super Chief, El Capitan,* and other members of its new streamliner fleet were brilliant silvery flashes trailing behind colorful EMD Es and Fs and

Dieselization

I love multi-unit consists of EMD F units. Unfortunately, the railroad I model—the Nickel Plate Road—gave F3s and F7s a good test, then ordered more Berkshires. They finally dieselized with Geeps and RS-3s.

Before I decided to give prototype modeling a try, I modeled the free-lanced Allegheny Midland. Its premise was that the Wheeling & Lake Erie, leased by the NKP in 1949, had built a line into West Virginia and across the ridge into Virginia to connect with fellow modeler Allen McClelland's free-lanced Virginian & Ohio.

When I finally could afford a roster of high-quality steam locomotives, I decided to shift the era backward in time to 1957. This date was the result of making a matrix showing key features of the NKP's Wheeling District year by year. (I made a similar table to choose a year for modeling the NKP's St. Louis line, **14**.) Since several of the Berkshires I had enjoyed watching as a kid in west-central Indiana had migrated to the Wheeling District in 1957, and the NKP had received SD9s from EMD and RSD-12s from Alco for use in southeastern Ohio (where the AM connected with the NKP) that year, the choice was easy.

Maybe the AM could also have some F units, I reasoned a bit too greedily. My daydreaming was cut short by EMD mechanical engineer Bill Darnaby, a friend from Purdue days. Bill pointed out that railroads that dieselized early with F units before the 1951 debut of the Geep were completely dieselized by the early '50s. "But what about the C&O?" I countered. "They had F units and Geeps, and steam lasted into 1956."

Nice try, but I lost that argument very quickly as Bill explained that the only reason steam lasted that long on the C&O and other large railroads was that they couldn't get diesels fast enough to retire steam as early as economics warranted. A smaller railroad like the Reading or Western Maryland—or Allegheny Midland—could dieselize much earlier as its power needs were met by the major builders.

Abandonments and mergers

The prosperity of the post-war years began to wane as the '50s matured.

16

Piggyback got its start in the West on the Southern Pacific between Los Angeles and San Francisco in 1953. By 1957, the railroad was moving more than 300 trailers each day, virtually all of them lettered for the SP or a subsidiary railroad. *Robert Hale*

In 1957, the New York, Ontario & Western, **15**, ended a long struggle that accompanied the decline of homes being heated with clean-burning anthracite coal. Even early dieselization with FTs and F3s, plus segments of Centralized Traffic Control, were inadequate to stem the flow of red ink. A steel strike and recession in 1958 slammed the door on the Nickel Plate's business plan to continue in steam until 1962; by the end of July 1958, even their vaunted Berkshires were cold.

Holdouts Norfolk & Western, **2-1**, Burlington's Colorado & Southern, the two big Canadian railroads, and a few smaller railroads segued into the 1960s still in steam. C&S 2-8-0 no. 641 was the last of the last, ending big-time steam's reign on U.S. railroads on October 11, 1962 (see *Classic Trains,* Spring 2017). The last Union Pacific regular run was July 1959 (but UP 4-8-4 no. 844 was never retired and continues in business and excursion service today).

From a modeling point of view, however, especially for prototype-based freelancers, modeling mainline steam past 1958 is reaching for it.

Trailers on flat cars

What we came to call piggyback service, or trailers on flat cars (TOFC), was just beginning in the early 1950s, but it caught on quickly. By January 1956, the Southern Pacific, which in 1953 had pioneered TOFC in the West, had extended it over more than 7,000 miles of its system from Portland, Ore., to Ogden, Utah, and New Orleans. The SP was moving almost 63,000 trailers annually by then, **16**, and—unlike what you see today—almost all of them were owned by SP or its subsidiaries. So piggyback trains of the 1950s will feature a high percentage of railroad-owned trailers.

The stars of the show

Colorful and innovative as the new diesels were, steam remained the star of the show during the transition era, as we'll explore in Chapter 2. Legions of fans drove countless miles, this before the Interstate highway system was complete, to witness steam's last hours. Those of us who witnessed at least some of it can re-create it on our model railroads, and those who missed the show entirely can enjoy convincing stand-ins on theirs.

611

CHAPTER TWO

Those black things with all the wheels

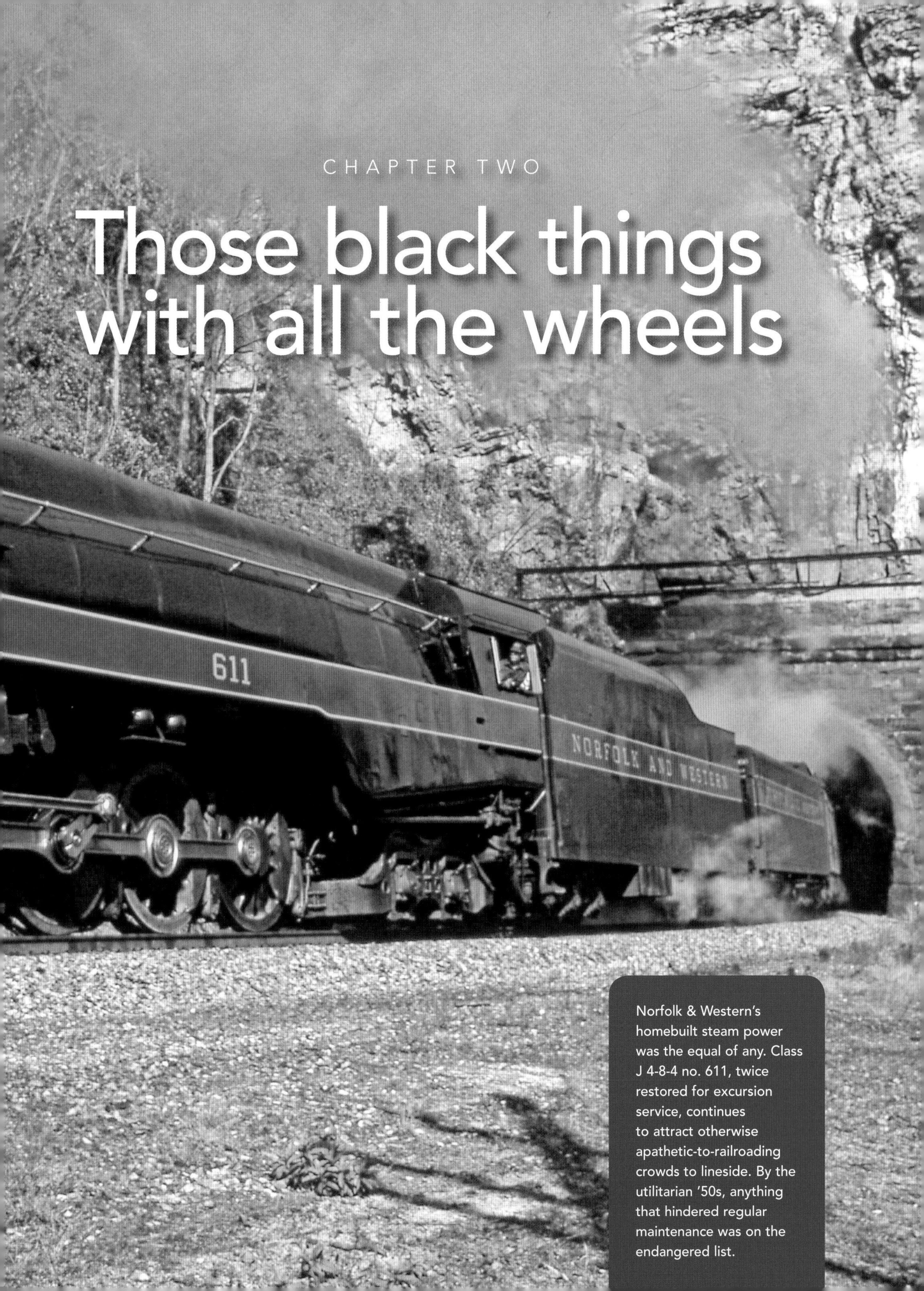

Norfolk & Western's homebuilt steam power was the equal of any. Class J 4-8-4 no. 611, twice restored for excursion service, continues to attract otherwise apathetic-to-railroading crowds to lineside. By the utilitarian '50s, anything that hindered regular maintenance was on the endangered list.

Steam as a method of railroad propulsion had a long run that began early in the 19th century. Some of the world's finest engineers did their level best to improve it and to ensure its longevity, **1**. That they were blindsided by Rudolf Diesel's invention is simply another case of technology moving forward, whether we fans and modelers of the Old Ways like it or not. We're simply finding enjoyment and modeling opportunity in re-creating the time when there was a clash of titans.

Steam's attributes and liabilities

We'll take a more detailed look at how to model steam power in the last full decade of its career in Chapter 6. But let's first set the stage for that encore performance by taking a close look at how a steam locomotive turned coal or oil and water into high-pressure steam and converted that to locomotion. What it took to support the steam locomotive is also of interest as we strive to do a credible job of modeling an engine servicing facility in the modern steam era, as we'll briefly review in Chapter 8.

The first steam locomotives were "saturated"—the steam was derived by heating boiler water surrounding the firebox. Tubes and flues carried heat and other products of combustion from the throat sheet of the firebox to the tube sheet that formed the forward end of the water vessel and the back end of the smokebox. Steam formed above the level of the water in the top of the boiler. The prominent steam dome atop the boiler was where saturated steam passed through a throttle valve, **2**, and then down into slide or piston valves, which alternately directed it to both the front and rear sides of the pistons.

As a locomotive accelerated, the engineer could cut off (shorten) the flow of steam to the pistons, thus allowing more time for it to expand, by carefully adjusting the reverse gear. On modern locomotives, this was done using a power reverse operated by a lever (Alco) or, more precisely, by a small wheel (Precision), **3**.

The strong draft needed to suck air into the firebox and through the tubes and flues was created when spent steam was exhausted from the pistons through ports in the valve chest and then into the smokebox via a nozzle that pointed upward under the smokestack.

Some modern locomotives had over-fire jets mounted on the sides of the firebox to inject even more oxygen into the firebox, **4**.

By the transition era, most locomotives were superheated. That is, the saturated steam from the initial application of heat to the boiler water was reheated ("superheated") by placing U-shaped tubes inside the large-diameter flues. Saturated steam was gathered in a superheater header box and directed back through the superheater tubes and then back forward to a front-end throttle, **5**. The hot gases from the firebox thus had direct contact with the superheater tubes, imparting a great deal of additional energy to the steam.

A superheated model locomotive should have an exhaust bark that's louder and sharper, and the throttle linkage should be connected externally to a front-end throttle rather than to one inside the steam dome.

Those domes

Atop every steam locomotive are almost always two or more domes. The most important is the steam dome, usually located at the highest point on the boiler, which is where saturated steam is collected. The forward (and sometimes rearmost) dome contains fine sand for traction.

As always, there's a notable exception to the rule: Look hard at the Virginian Berkshire at the top of **6**. At first glance it looks a lot like a Chesapeake & Ohio 2-8-4 (Kanawha) below it, upon which its design was based, with one notable exception: Where's the steam dome?

It's hidden inside the sand dome, but why? The C&O 2-8-4s had the steam dome where one would expect it at the boiler's highest point, which caused the massive sand dome to be moved forward, giving the locomotive a somewhat front-end-heavy appearance. The C&O 2-10-4 and its scaled-down siblings, the Nickel Plate and Pere Marquette 2-8-4s plus the near-copy Richmond, Fredericksburg & Potomac 2-8-4s, achieved a more balanced look by moving the steam dome forward.

2

A saturated steam locomotive had a throttle on the steam dome that allowed the engineer to regulate the flow of steam from the boiler to the valves and into the cylinders. *Lima Locomotive Works*

3

As the valve gear and associated levers and links got heavier, power reverse gear was employed. The Alco-type reverse cylinder (left) was operated by a lever mounted to the right of the engineer; the more compact Precision reverse gear (right) was adjusted with a small wheel. *Left: Tony Koester; Right: Willis McCaleb, Jay Williams collection*

Over-fire jets, as mounted on the firebox of this C&O Allegheny, injected more air into the firebox, improving the combustion and delivering more work for a given quantity of coal. *Trains magazine collection*

Superheated locomotives routed saturated steam back through the boiler flues via U-shaped superheater tubes to add even more energy to the steam. The throttle was therefore located near the front end of the boiler. The added energy was noticeable with each blast of the exhaust. *Willis McCaleb*

6

Virginian's Lima-built near-copies (top) of C&O 2-8-4s (middle) had larger tenders and one other distinctive visual attribute: The steam dome was hidden inside the sand dome. Whether this was done for aesthetic or balance reasons or to help keep the traction sand dry is unclear. NKP (bottom) and Pere Marquette Berkshires had the large sand dome at the boiler apex, resulting in a more balanced appearance. *Three photos: Lima Locomotive Works*

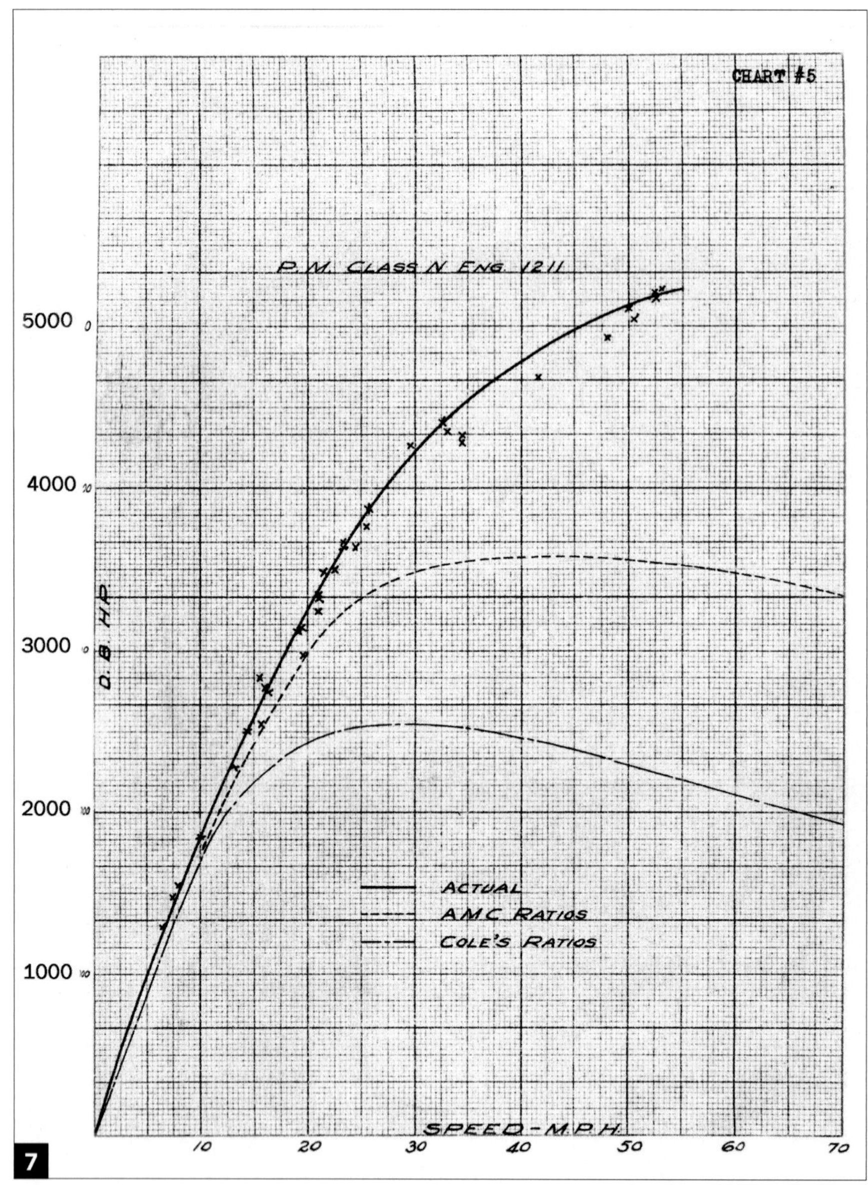

This graph made during tests of Pere Marquette N-class Berkshire 1211 shows that horsepower measured at the drawbar (vertical axis) increased with speed, topping 5,000 hp at 50 mph. Once a Super Power locomotive got its train started, it could move it right along. *Courtesy Karen Parker and the C&O Historical Society*

The Virginian's semi-clones of the C&O engines restored the steam dome to its logical place but retained a balanced look by covering it with the sand dome. Did this help keep the sand dry? Or was it a purely a matter of aesthetics? We don't know.

In any event, those 2-8-4s were powerful locomotives. A dynamometer test of a Pere Marquette Berkshire proved that they could develop upwards of 5,000 horsepower at speed, shown in the chart in **7**.

The use of the same basic design on at least five railroads (including the Wheeling & Lake Erie, which was leased by the NKP in 1949), and its availability from many manufacturers and importers in all popular scales from N to large scale, suggests that this Advisory Mechanical Committee design would be a good choice for a freelanced east-of-the-Mississippi railroad.

Engine or locomotive?

You may have noticed my occasional shift in terminology from "locomotive" to "engine," and the terms are often used interchangeably. Technically, the engine is the driving gear located under the boiler; an articulated locomotive such as a 2-6-6-2 has two engines.

And articulateds came in two flavors: simple and compound. A compound locomotive uses steam twice, first in the rear high-pressure cylinders and then again in the much larger low-pressure cylinders up front, **8**. Compounds were often called Mallets in honor of the French originator of the concept, Anatole Mallet ("mal-*lay*"). Americans pronounce the word "malley" (not "mallet"). The exhaust is the same as a conventional steam locomotive: four blasts per driver revolution, an important consideration when choosing a sound decoder.

Most modern articulateds like the Union Pacific's famous 4-6-6-4 Challengers, **9**, and 4-8-8-4 Big Boys were simple: Both sets of cylinders received high-pressure superheated steam. And since both cylinders exhausted directly to the atmosphere, there were four blasts per driver revolution times two, usually through a pair of smokestacks. This could lead to a syncopated beat as the front and rear drivers—engines—shifted in and out of synchronization. This is also an important distinction when choosing and applying a sound decoder.

If a steam locomotive didn't move for any length of time, steam vapor could condense into water in the cylinders. To avoid blowing off the cylinder heads as the pistons moved forward and aft, the engineer would open cylinder cocks that allowed the condensation to be blown overboard. Newer sound decoders have this feature, which makes a loud, syncopated roar of escaping steam for several revolutions of the drivers.

The holdouts

By 1955, it was clear to most railroads that steam power was on its way out. But a few major railroads felt that the old beasts still had some life left in them and kept mainline steam active until 1960.

A well-known holdout was the Norfolk & Western, which built much of its own steam power in its Roanoke, Va., Shops, **1**, including 0-8-0 no.

Compound articulateds are easy to spot in that the rear high-pressure cylinders are much smaller than the front low-pressure cylinders, which reuse steam exhausted from the rear cylinders. They have a single exhaust stack (at the front) and have a "normal," if slightly softer, exhaust beat. *Trains magazine collection*

Simple articulateds have four standard high-pressure cylinders; UP's Challengers are examples. Steam is used only once, so they have twin exhaust stacks and emit two distinct exhaust blasts that drift in and out of synchronization. *Mike Brock*

244 in December 1953, the last steam locomotive built for a U.S. railroad. Another holdout was the Duluth, Missabe & Iron Range, **10**. Frank King's book *Locomotives of the Duluth, Missabe & Iron Range* lists the last mainline steam run as Yellowstone 2-8-8-4 no. 222 on July 5, 1960. Few steam runs occurred in 1959, but an up-tick in ore traffic brought some back out in 1960.

Several examples in several classes of steam (including three of the 18 Yellowstones) were not officially retired or disposed of until 1963, with one holdout to 1967. One of the 1963 retirements was no. 225, donated to the Proctor Development Council. Number 227 went to Duluth's Lake Superior Railroad Museum, and 229 went to the Lake County Historical Society in Two Harbors in 1967.

Burlington subsidiary Colorado & Southern was another railroad that operated big steam into 1960. The Nickel Plate's plans to continue steam operation until 1962 fell

Duluth, Missabe & Iron Range moved countless tons of iron ore behind 18 massive 2-8-8-4 Yellowstones, three of which have been preserved. Here 234 in fresh paint (top) pauses by one of the washer/concentrator plants. The last run was by No. 222 in July 1960. The Oliver Mining EMD cow-calf switcher above is spotting carloads of raw ore that will be concentrated for shipment to ore docks and lake boats. No. 228 was photographed at Biwabik (middle). Preserved No. 225 (bottom) is slowly sinking into the turf at Proctor; the other two are under cover.

Top two photos courtesy Missabe Railroad Historical Society

10

Even lowly 0-6-0 and 0-8-0 switch engines designed to USRA specs were handsome and long-lasting machines. The ubiquitous USRA 0-6-0 and 0-8-0 switchers (Proto 2000 by Walthers) are ideal choices for a freelanced railroad.

One of the most handsome USRA designs was the light Mikado (2-8-2). It and its heavier siblings were found on railroads across the nation. Many were delivered with tenders of modest capacity to fit short turntables, but their looks markedly improved with larger tenders (see **6-6**). NKP H-6o 587 is a Powerhouse model detailed, painted, and weathered by Mark Guiffre and Ken Dasaro.

through during the 1958 recession and steel strike; some 0-8-0s were fired up in 1959. According to Lloyd Stagner's *Union Pacific Motive Power in Transition,* the UP closed out regular steam operations when engines 3713 and 3703 made runs during the evening of July 22, 1959, with 3713 arriving at Cheyenne, Wyo., and 3703 arriving at North Platte, Neb. Excursions pulled by 4-8-4 no. 844 continue to this day.

Excursions present opportunities for modelers to run fan trips well beyond the transition era—but not a plausible excuse to freely mix steam and diesel beyond 1960.

The USRA legacy

We tend to think in terms of mass-produced locomotives that differ only in minor details as products of the diesel era. But Baldwin catalogued a lot of similar steam locomotives, especially for short lines.

One of the more prominent examples of standardized locomotives resulted from the government takeover of U.S. railroads during World War I. The United States Railway Administration (USRA) developed standard designs for both locomotives and rolling stock. Perhaps surprisingly, almost all of them were excellent. Even the railroads that resisted accepting USRA locomotive designs at the time wound up ordering more after the war.

Among the better-known designs are the 0-6-0 and 0-8-0 switchers, **11**; the light and heavy Mikados, **12**, and Pacifics; and 2-6-6-2 and 2-8-8-2 compound articulateds. This predated the advent of the Berkshire wheel arrangement, but the Advisory Mechanical Committee-designed NKP-PM-C&O-RF&P-VGN 2-8-4s bear strong evidence of USRA influence.

Using one of the USRA locomotive types as the basis for a freelanced roster is therefore a very plausible approach, as we'll discuss in Chapter 6.

405
05
405
GREEN
MOUNTAIN
R.R.

CHAPTER THREE

The giant killers

How it all began: Former Rutland RS-1 No. 405 chirps into Chester, Vt., on the Green Mountain's portion of the former RUT main line between Bellows Falls, N.H., and Rutland, Vt. At a time when enclosed cab units were all the rage, this seemingly unremarkable machine set the trend for locomotive design that continues to this day.

An A-B set of drawbar-connected Boston & Maine FTs (in Boston in March 1950) shows the distinctive four-porthole styling and close spacing between units. The boxy structure on the roof is the dynamic brake housing. The F unit's "bulldog" carbody styling blended nicely with passenger cars but did not facilitate clear views to the rear for en route freight-car switching. *George Corey*

The blunt-nosed Alco FA was a dramatic improvement over earlier Alco cab-unit designs. Western Maryland 301 was one of four FA-2s on its roster. The vertical shutters on FA-1s are mounted farther back near the rear bulkhead.

EMD's BL1 (only one was built) and BL2 were intended to provide greater visibility for switching branch lines. The rapidly growing popularity of the road switcher brought production to an end after 13 months. Western Maryland's two BL2s were later paired with "slugs" for switching in Hagerstown, Md.

What began as boxy urban switchers and progressed to a more streamlined carbody segued to the ultimate diesel locomotive, the road switcher. Before that, however, styling and streamlining were primary considerations shortly before and, more dramatically, after World War II. Electro-Motive Corporation and successor Electro-Motive Division (of GM) E and F units proved to be an apt canvas for myriad colorful liveries. Alco was able to join the party following the war with its elegant PAs and FAs, as did Baldwin and finally Fairbanks-Morse with distinctively styled cab and hood units. It was quite a time!

The wagon train

A seemingly endless stream of covered wagons headed west as the remainder of what became the 48 contiguous United States was settled in the 1800s. An equally irresistible and revolutionary stream of "covered wagons," as diesel-electric cab units came to be called, spread out across the continent roughly a century later. No matter how hard steam enthusiasts, and not a few professional railroaders, bemoaned this incoming tsunami, it was not to be slowed, let alone stopped.

Although, like any large, heavy piece of machinery, a diesel locomotive requires periodic maintenance, that work was far less labor intensive and frequent than the routine servicing of even the best modern steam locomotive. True, some railroads ran their modern steam power over multiple divisions, but it was more typical to cut the engine off at each division point for inspection and servicing, as we discussed in the previous chapter. Diesels, on the other hand, were more like the family car, almost a turn-key machine.

5

Alco followed the RS-1 with the 1,500-h.p. RS-2 (1946–50) and similar 1,600-hp RS-3 (1950–56); six-motor (RSD-4 and -5) and A1A (RSC-2 and -3) versions were also offered. WM 186 is an RS-3: Note boxes on frame fore and aft of the cab.

From switcher to road switcher

The diesel-electric locomotive first gained a foothold in the form of switch engines. Here the railroads had the opportunity to discover that a relatively low-power, diminutive diesel could start and switch a train that took a 4,000-hp modern steam locomotive to bring into the yard. They also discovered that the diesel switcher spent a lot less time in the roundhouse or shops than its steam contemporaries.

By stretching the frame a bit, adding a short hood, and equipping a switcher with better-riding road trucks, the road switcher was born. Compare an Alco S-class switcher

6

Dick Dilworth's utilitarian "Geep" debuted in late 1949. The carbody remained largely unchanged from the first GP7 until the introduction of the GP30 in 1961. Nickel Plate 490 is an early model GP9; note the four 36"-diameter rooftop fans (later GP9s had two 48"-diameter fans). GP7s also had two sets of louvers on the panels below the cab number. The NKP eschewed dynamic brakes. The unit has been repainted in the railroad's post-1959 wide-stripe scheme.

7

Like Alco and Baldwin, EMD offered its hood units in stretched six-motor versions called SDs. Bessemer & Lake Erie 828 is an SD9, as evidenced by the classification lights being outboard of the number boards (as they also are on SD18s).

to the groundbreaking—indeed, revolutionary!—RS-1, **1**.

Another discovery was that the typical diesel was a jack of all trades. Steam power tended to be built to do a specific job. Lacking a lead (pony) truck, for example, an 0-6-0 or 0-8-0 switcher was an uncomfortable ride at whatever top speed its small-diameter drivers could muster. But a diesel switcher could be factory-equipped with cables that allowed it to operate in multi-unit consists—the Lehigh Valley's SW8 "Pups" are good examples—and used out on the high iron. And the road switcher was equally at home in the yard and main line.

That was not true of the first road cab units such as the EMD Fs, **2**, and Alco FAs, **3**. The rearward visibility was so restricted that when EMD confidently painted a pair of F7 A-unit demonstrators in a paint scheme inspired by the Nickel Plate Road's Alco PA-1 passenger units, the NKP sent them packing and instead ordered another ten Berkshires, this in 1949. The rearward visibility was simply too limited, especially when used on trains that did a lot of switching between terminals.

But on the main line, the cab units did just fine—until the Geep came along. EMD had heard the complaints about restricted visibility for switching moves, and the result was the BL1 demonstrator and then the BL2, **4**. The BL designation showed they had high hopes for its use as a branchline locomotive. The BL1 demo unit didn't even have multiple-unit hoses and cables. (It was rebuilt into a BL2 and sold to the Chicago & Eastern Illinois along with two more BL2s.)

But this homely "chain drive" unit—a moniker it earned by virtue of its sloping sheet metal on either side of the cab that looked like a bicycle's chain guard (I always thought they looked like basset hounds)—was expensive to build and never caught on.

I grew up along a Rock Island line in Iowa; the Rock had BL2s. We moved to Cayuga, Indiana, and the Chicago & Eastern Illinois had the original BL1 and two BL2s, one of which hosted my second-ever cab ride. In 1958, we moved to Long Beach near Michigan City, Ind., and I often got rides in the Monon's BL2s. In 1969, we moved to New Jersey, and I made railfan jaunts north to New England, including to Maine's Bangor & Aroostook, which had BL2s. I also made regular trips south to photograph the Western Maryland, which had a pair of BL2s. From my perspective, the BL2 seemed to be a bestseller!

Alco had shown the way forward with the RS-1 way back in 1941 and later with the much-improved, more-powerful RS-2 (1946) and RS-3 (1950), **5**. Clearly, EMD had some catching up to do, and engineer Dick Dilworth led the way with the GP7 (1949), **6**. He apparently felt that the Geep (pronounced "jeep") as it became known, was both needed to compete yet too ugly to disrupt the sales of the still-popular F units. Right and wrong, respectively: Between October 1949 and August 1963, EMD's U.S. and Canadian plants produced more than 2,700 GP7s and more than 4,000 GP9s for the U.S., Canada, and Mexico.

But the FT and its younger brothers did quite well too. EMC and EMD sold more than 7,000 FT, F2, F3, F7, and F9 units between November 1939 and April 1957. Still, the Geep and its six-motor companion, the SD (Special Duty, **7**), eventually won the day.

8

Trucks with two powered axles, like those on Monon FM H15-44 no. 45 (top; re-engined by EMD), are designated B-B. Three powered axles, as on B&LE Baldwin DRS-6-6-15 no. 405, middle, are C-C; and those with an idler center axle (passenger and some branchline units), as on Erie Lackawanna E8 No. 831, bottom, are A1A-A1A.

Terminology

Before we venture too much farther into Diesel Land, let's be sure the terms we use are understood.

Diesel trucks are designated by noting which axles are powered and which are along for the ride. A switcher or road switcher typically has two four-axle trucks, each axle being powered by its own traction motor, which in turn is powered by a generator hooked to one end of a diesel prime mover. Such a truck is designated by the letter B, so the switcher or road switcher is a B-B unit, **8**. If each truck has three powered axles, it's a C-C unit. ("D" trucks with four powered axles such as Union Pacific's EMD DDA40X monsters were not part of the Transition Era except on electrics, nor were diesel locomotives that had hydraulic drives.) Passenger locomotives such as the EMD E unit or Alco PA ride on trucks with the two outer axles powered and an unpowered center axle to smooth out the ride. These trucks are therefore designated A1A.

EMD boosted their engines' air intake by adding a direct-drive blower (supercharger). Alco chose to compress the intake air by inserting a turbo-charger into the exhaust stream. The lag between adding fuel and the turbo spinning up to provide the oxygen to burn it is what caused the plume of smoke, especially as an idling Alco was suddenly called into service. With any luck at all, we won't emulate this effect with our models.

To further boost horsepower, EMD added a turbocharger to their 567-series prime movers (the number represents the cubic inches of each cylinder of the engine) beginning with the SD24 in July 1958 and the GP20 in December 1959. The traditional pair of exhaust stacks thus gave way to a single large turbo stack behind the cab, **9**. (The 645-series prime mover was introduced in 1965.)

To help you choose the proper sound decoder, note that Alco designated their diesel engines (prime movers) by the month and year the design was introduced. The switchers, RS-1s, and DL-109s had 539 engines, either normally aspirated or with turbos that sounded a lot like jet engines spooling up, hence the 539 and 539T designations. Alco then developed the turbocharged 244 engine, which powered the RS-2 and RS-3 through the RSD-7, the FA-1 and -2, and all PAs.

Alco replaced the troublesome 244 with the 251, which powered FPA-4s, RS-11s, RS-32s, RS-36s, RSD-12s, RSD-15s, and all Century series locomotives.

Transitioning

Ever drive a stick-shift car? That is similar to the way an engineer had to operate early and some later-model diesel-electric locomotives. It's that

9

When EMD turbocharged its 567-series prime mover, resulting in the 2,000-h.p. GP20 (shown) and 2,400-h.p. SD24, the exhaust had to be routed through a single large stack (on the roof between the forward fan and dynamic brake housing). That practice continues today, with non-turbo units such as the GP38-2 having the traditional pair of exhaust stacks. *EMD*

Diesel transitioning

During a discussion as to which units still had manual transition, as opposed to automatic (which was originally an extra-cost option), Chicago & Eastern Illinois modeler Edwin Cooper cited EMD Locomotive Reference Data from January 1, 1959. This lists what type of transition C&EI units were built with: automatic, manual, or both (A&M).

- NW2: Automatic
- SW7: Automatic
- E7A: Automatic
- BL1: A&M
- BL2: A&M
- GP7: Automatic
- GP9: Automatic
- F3A 1200-1203, 1400-1403: Manual
- F3B 1300-1301, 1500-1501: Manual
- F3A 1400-1409, 1204-1205: A&M
- F3B 1502-1504: A&M
- FP7: A&M

hyphen between diesel and electric that explains why this happened.

Almost all "diesel" locomotives are in fact diesel-electrics. The diesel engine drives a d.c. generator (or an a.c. alternator in most modern locomotives). The electricity from the generator is sent to d.c. motors mounted on the driving axles. As the locomotive starts to move, the voltage from the generator is split evenly between the four or six motors—that is, they're wired in series. As it gains speed, the engine drops to idle and relays switch the circuitry so that pairs of motors are in parallel, thus receiving half voltage. As the speed continues to build, the engine again drops to idle, and the motors are then switched into parallel, each motor receiving full voltage from the generator.

This has direct implications for the modeler. As a train gains speed, you should expect the sound decoder to replicate these transitioning steps, with two distinct sound pauses where the prime mover drops to idle without decreasing speed. Moreover, since some early cab and hood units had manual transition controlled by the engineer, you may want that feature on your decoders (see "Diesel transitioning" at left).

Some newer decoders such as the LokSound (pronounced "loke-sound") from ESU have a "power hold" feature that allows you to set a constant speed as a heavy train ascends a grade while the throttle continues to be notched up. Most diesels had eight-notch throttles, so even without transitioning there can be distinct engine sound changes.

Dynamic brakes were a feature even on the FT. This allowed the engineer to convert the motors to generators to provide resistance, and the current generated was set to banks of resistors in the roof. Fans dissipated the heat.

The presence of dynamic brakes is easy to discern by looking at the top of the carbody or hood, **10**. Also refer to the discussion on dynamic brakes in Chapter 7 about modeling diesels.

Familiar-looking paint schemes

We'll discuss the modeling implications in more detail in Chapter 7, but for now let's acknowledge that it's no coincidence that the paint schemes applied to many railroads' diesel fleets, especially those from EMD, looked similar, **11**. That's because the builder's small cadre of graphic-design artists produced most of those schemes in-house. Jim Boyd, who worked as a field technician for EMD in the 1960s, visited the art studio and found the French curves with those distinctive, oft-repeated arcs marked on them!

You can also discern standard practices by graphic designers who worked for Alco and Baldwin. But even many of their products bore the mark of EMD stylists, simply because EMD got there first. During WWII when diesels were first entering the market in force, the War Production Board restricted production of road units to Electro-Motive; Alco and Baldwin could build only switchers.

EMD usually drew a bright line between paint schemes applied to cab units and those typically used for road switchers (Geeps and SDs) or yard switchers. But there were overlaps: Consider the similar black-and-white livery applied to Chicago & Eastern Illinois Geeps and SWs, **12**. That's not much of a stretch when you think about the evolution of switchers into road switchers.

What's too new?

I assume you're interested in faithfully modeling some aspect of the transition era, as you bought this book. That brings good news and bad news: It places you solidly inside the most popular era to model, which means it is well served by model manufacturers and importers. The bad news is that it is quite obvious when one strays even a year or two beyond that era, as locomotive styling changed rather dramatically.

Consider the ubiquitous Geep, for example. It takes a bit of an expert to tell a GP7 from an early GP9; some louvers or the lack thereof below the cab are the chief spotting feature. (Both were available with or without the dynamic brake "blister.") By 1960, the end year for this era, EMD had upped the horsepower by a whopping 50 to create the GP18, which in most respects looked a lot like a late-model GP9 with two 48" instead of four 36" roof-mounted exhaust fans.

Then came dramatic change: Next up in 1961 was the GP30 with its distinctive, one-off "automotive" styling. There's no way to sneak a GP30 onto a transition-era railroad.

Things aren't much better for Alco fans. The rounded profiles of the handsomely drawn RS-2 and RS-3 segued into the boxy RSD-7, **13**, in 1954 and the RS-11 in 1956. Next up in 1961 was the RS-32 and look-alike RS-36, which debuted in 1962.

10

Diesels with dynamic brakes usually have external evidence of their installation. EMD FTs, top, have boxy structures on the roof. Geeps and SDs had prominent "blisters" centered on the long hood, middle. The RS-11, bottom, and later Alcos had rectangular panels on the roof and five (instead of three) intake grilles on the top of each side. *B&M FT: George Corey; B&O Geeps: Tony Koester; RS-11: Alco*

11

It's no coincidence that seemingly different paint schemes embraced the same basic graphic elements. EMD's styling department used French curves to duplicate the nose striping on quite a few railroads' Es and Fs. *D&RGW: L.O. Merrill; B&M: George Corey*

12

Hood unit paint schemes were often quite different from cab units, and switchers sometimes had yet another scheme. The Chicago & Eastern Illinois used blue and orange on cab units, but they adopted switcher-style striping for their Geeps (here a GP35 at Cayuga, Ind.).

13

The smoothly rounded RS-2 and RS-3 carbodies gave way to a taller, boxy profile when the six-motor RSD-7 (including B&LE 882) debuted in 1954 and the four-motor RS-11 debuted in 1956. Alco's improved 251-series prime mover also replaced the 244.

They usually arrived with increasingly popular low short hoods. Although they resemble the RS-11, populating a transition-era layout with -32s or -36s would be an obvious anachronism.

Baldwin merged with Lima and Hamilton in 1950 to form B-L-H and exited the business in 1956. Any of its offerings are therefore fair game.

Fairbanks-Morse, provider of powerful opposed-piston diesel engines to the U.S. Navy and commercial users, exited the locomotive-building market three years after the end of the transition era. Virtually all of their switchers and passenger units, and both B-B and C-C hood units, are therefore suitable for use on a transition-era model railroad.

Utilization

As diesel bumped steam from various assignments, it was understandably easy to think in terms of replacing *this* with *that.* But diesels didn't need to be laid up overnight for maintenance.

The bankers who loaned the railroads the money to acquire new fleets of diesel-electrics observed this lack of utilization and insisted that these assets be better utilized. No sense having the power for a three-days-a-week local freight or daily except Saturday and Sunday local passenger run idling away when the unit could be consisted with others for road duty on its off days.

Passenger units

Diesels designated for powering passenger trains needed boilers (steam generators) and water tanks to provide the steam that operated both heating and cooling systems on even the most modern lightweight passenger cars. The presence of a boiler resulted in air intakes and exhaust stacks. On cab units, these appliances were located at the rear of the carbody; on hood units, the short hood provided space, **14**.

Back in the early days of plastic diesel models, manufacturers typically included steam-generator details whether or not the prototype for which they were painted and numbered had them. Today's models are usually detailed to match their prototypes.

14

Steam generators on F units, top, were located below the rear roof hatch. On hood units, middle, it was accommodated in a high short hood. On Geeps, air reservoirs were moved to roof to provide space for a water tank under the frame adjacent to the fuel tank, as on this Maumee "torpedo-boat" GP9 modeled by Bill Darnaby.
ATSF: Don Sims; NKP: Tony Koester; Maumee: Bill Darnaby

CHAPTER FOUR

Electrified railroading

The self-proclaimed Standard Railroad of the World, the Pennsylvania, thought well outside the box when creating the timeless GG1. Equally at home in freight or passenger service, the GG1 performed its duties between New York City, Washington, D.C., and Harrisburg, Pa., from 1935 until the last few were retired in 1983. Three were painted silver in the 1950s, but, as Andy Rubbo reports, rusty water running down from the steel pantograph shoes made it hard to keep them clean. Here an express comprising Budd's "Keystone" cars behind 4880 meets an owl-eyed MP54 local on Andy's HO railroad.
Andy Rubbo

Unlike Europe, mainline electrification never really caught on in North America. There were notable exceptions, however, and we'll take a brief look at several of those that survived into the transition era. A few interurban railways with electric freight components also lasted into and beyond the 1950s. Despite the added complexity of modeling the overhead-wire structure, whether it's powered or just for looks, it's the most faithful approach to small-scale modeling: totally electric!

2

Pennsylvania's GE 4,400-hp. a.c./d.c. rectifier E44, shown here on Andy Rubbo's HO tribute to the PRR's (and Penn Central's) New York–Washington main line, was delivered beginning in 1960. They replaced the old P5 freight motors, which ran side-by-side with the E44 until their retirement in the early '60s. Virginian's earlier E33 rectifier from GE has essentially the same wheelbase and trucks. The early piggyback train features the Walthers PRR F39 flat cars, constructed in 1955 specifically for this service. They became Trailer Train's first cars when TT began operations in 1956. *Andy Rubbo*

Mainline electrification

The story of big-time electrification is well covered in William D. Middleton's *When the Steam Railroads Electrified* (Kalmbach, 1974). My goal here is merely to relate which of those massive projects managed to survive into or beyond the transition era. If the idea of modeling mainline electrification catches your fancy, reading Middleton's book will be an extremely rewarding rail-history experience.

We will ignore primarily commuter operations under wire such as the Illinois Central lines south of Chicago, the Reading's, and the Long Island's.

Prototype modelers can easily learn whether their favorite railroad operated a specific type of electric locomotive in a given year, so the following information will probably be of more interest and value to the prototype-based freelancer.

Transition-era electrics

Of the big-time railroads that were still partially electrified in the transition era, only the former Pennsylvania main between New York and Washington, D.C., **1** and **2**, and New Haven's main between New York and Boston—the Northeast Corridor—continue under catenary today. Although the overhead now powers only passenger trains, freight service continued under wire well past the transition era.

Fortunately, the era of the PRR's incomparable GG1s continues on scale model railroads cross the continent. Andy Rubbo, who models what is now considered the Northeast Corridor (which runs through his backyard!), summarizes the options: "Broadway Limited (and later MTH and soon BLI again) made GG1s with sound decoders. The E44 was produced in brass by Alco Models and Alpha Models, but that was some 40 years ago. The prototype E44s began showing up on the PRR in 1960, so they barely fit into a transition-era layout."

He notes that the ex-Virginian EL-C (Penn Central E33) was produced in brass (by Overland and Alco) and in plastic by Bachmann. "They never ran in PRR days," he adds, "but rather came to Penn Central when the New Haven was absorbed in 1969. BLI is doing a PRR P5 boxcab electric, which could be considered a prime example of that timeframe. They saw service from the 1930s to the early '60s."

The New Haven was still using its own electrics, **3**, and the former Virginian road-switcher-like EL-Cs when it merged into Conrail in 1980.

The idea that electrics are relatively silent is disabused by HO models of electrics with DCC sound systems.

3

New Haven ran electrified freight trains into the Penn Central era. Here boxcab and streamlined electric motors team up at New Haven, Conn., on what is now Amtrak's busy Northeast Corridor. *A.S. Arnold*

They make more noise than you might think! During the 1960s, I often watched the Chicago South Shore & South Bend's three big 800-series "Little Joe" electrics at work, **4**, and the cooling fans made the same sound whether they were coasting along "light" or pulling a sizable string of loaded coal hoppers. They never seemed to work up a sweat.

Appalachian electrification

Smoke problems in tunnels and the slow pace of big articulated steam locomotives hauling coal from the central Appalachians to tidewater caused the Norfolk & Western to electrify the grade through Elkhorn Tunnel in West Virginia in the early 1900s, **5**. Shortly after WWII, however, a five-mile line relocation and double-track tunnel through Elkhorn Mountain reduced the eastbound (loaded-car) grade from 2 to 1.4 percent. This and the N&W's fleet of modern Y-class articulateds made the electrification unnecessary. After the new line opened in 1950, the big boxcabs were out of work.

4

The South Shore acquired three of the 20 streamlined GE motors intended for Stalinist Russia. The trio basks in the sun under the Roeske Avenue overpass at the Shops in Michigan City, Ind. Milwaukee Road got 12 and five went to Brazil.

The Virginian electrified the 34 miles from Mullens and Elmore, W.Va., up through Clark's Gap to Princeton, and then on to Roanoke, Va., with boxcabs resembling those employed by the N&W, **6**. Those were joined in 1948 by handsomely streamlined twin-unit GE EL-2Bs that together produced 6,800 hp. They were joined in 1956 and '57 by a dozen 3,000-h.p. GE EL-C rectifier hood units, perhaps better known by the E33 designation applied by Penn Central.

The Virginian's electrification survived the 1959 merger into the Norfolk & Western, but just barely. The merger allowed heavy coal drags to be routed over the N&W's easier gradient to Norfolk, thus avoiding the notorious climb up to Clark's Gap. The power from the big coal-fired power-generation plant at the Narrows was shut off in June 1962.

Virginian modeler Gerry Albers (see *Model Railroad Planning* 2005 and *Great Model Railroads* 2014) acquired Overland brass EL-Cs (E33s) and Global Imports EL-2B streamliners. He equipped them with LokSound decoders. Since then, SoundTraxx has come out with electric-locomotive sound decoders: Econami 883005 and Tsunami2 886001.

As of early 2017, specific trolley and electric-locomotive sound decoders were under development for ESU's LokSound line. "We will be doing

Norfolk & Western was a heavy user of big articulated steam locomotives but electrified the grade through Elkhorn Tunnel in West Virginia. The main was relocated following World War II and electrification discontinued. Here motor 2503 climbs the 2 percent grade near Maybeury, W.Va., on June 12, 1950. *Norfolk & Western*

The Virginian, primarily a coal hauler to tidewater, employed "Squarehead" electrics, upper left, in one- to three-unit sets for the climb to Clarks Gap. The clank-clank of their side rods was a prominent source of noise from the otherwise quiet electrics. Virginian EL-3A number 107 is at Tralee Collieries, just east of Elmore Yard, W.Va., on June 12, 1950. Virginian's massive streamlined electrics, classified EL-2B, upper right, arrived from GE in 1948, and 128 is at Roanoke, Va., in January 1957. A dozen hood-unit-style, 3,000-hp EL-C rectified motors were delivered by GE in 1956-57; nos. 135 and 134 are entering the Roanoke, Va., yard in May 1957. *Squareheads: Richard Cook, Aubrey Wiley collection; EL-2B: Aubrey Wiley; EL-C: Tony Koester*

The Boston & Maine eliminated the steam-exhaust problem in Hoosac Tunnel with electrification. It survived until just after WWII when dieselization came to the rescue. *Trains magazine collection*

Until 1953, the downtown Cleveland Union Terminal was electrified with boxcabs built by GE and Alco to deal with smoke abatement. CUT motors handled both NYC and NKP passenger trains. *R.V. Nixon*

The Milwaukee Road had two electrified districts: Harlowton, Mont., to Avery, Idaho, and Othello to Tacoma, Wash. Boxcabs, bipolars, and Little Joes worked the line. Electrification ended in the early 1970s. *Trains magazine collection*

everything from old trolley cars up through the E33s, Little Joes, and GG1s," reports ESU's Matt Herman. "That will continue through the most modern electrics and dual modes being used today."

Gerry also has two sets of three-unit Alco Models and Overland boxcab EL-3A "Squareheads" which present an interesting sound challenge: They require the usual electric cooling-fan sounds plus the side-rod clank typical of steam locomotives. Using a steam decoder with everything but the rod clank turned off should do the trick.

In the heavily eroded but still rugged northeastern Appalachians, in 1911 the Boston & Maine electrified the 25,000-foot Hoosac Tunnel in northwestern Massachusetts, **7**, another example of the poor ventilation inside long tunnels causing severe problems for man and machine alike. The B&M's early embrace of the diesel-electric ended electrification in 1946.

Midwestern electrification

The Van Sweringen brothers' massive Cleveland Union Terminal project employed 22 GE and Alco 2-C+C-2 electric locomotives, **8**, to move New York Central passenger trains 17 miles from Collinwood on the east side of Cleveland to Linndale on the west side. Nickel Plate passenger trains were electrified by CUT motors five miles from East 40th Street to West 38th Street. When electrification ended in 1953, those big motors were moved to parent New York Central and rewired for 600-volt DC third-rail operation out of New York City. Some operated into the 1970s.

Electric locomotives were used until 1953 to move tonnage through the Detroit River Tunnel between Detroit, Mich., and Windsor, Ontario over New York Central System's Michigan Central. Canadian National subsidiary Grand Trunk, a late converter to diesels, ended electrification through the nearby St. Clair Tunnel in 1958.

Electrification in the West

My greatest regret as a railfan is never making the trek west to witness either of the Milwaukee Road's fabled Pacific

10

The Butte, Anaconda & Pacific had 17 GE boxcabs that in 1957 were augmented with a pair of GE road-switcher-like units. Electrification ended in 1967. This photo shows a BA&P ore train crossing over an electrified Milwaukee Road freight; the other track is Northern Pacific. *Trains magazine collection*

11

Great Northern coped with smoke in Cascade Tunnel with Baldwin-Westinghouse and GE boxcabs and streamlined electrics from GE. One Y-1 boxcab, 5011, was damaged and rebuilt with EMD F unit cabs. Electrification ended in 1956. *Robert F. Collins*

12

The Mason City & Clear Lake survives today as Iowa Traction using steeple-cab motors that date back to Prohibition days.

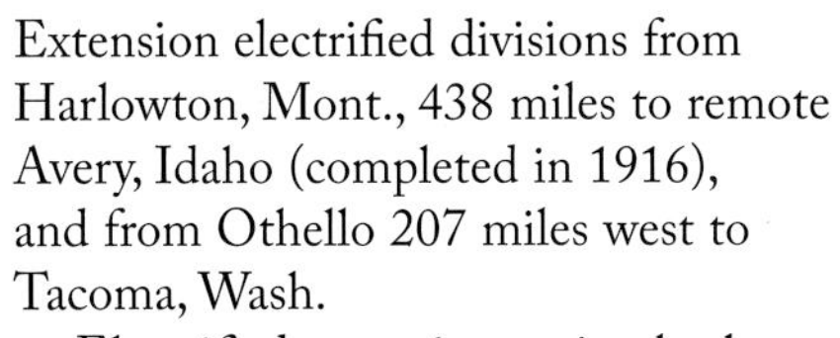

Extension electrified divisions from Harlowton, Mont., 438 miles to remote Avery, Idaho (completed in 1916), and from Othello 207 miles west to Tacoma, Wash.

Electrified operations using both Milwaukee's 12 Little Joes and the aging boxcabs, **9**, lasted well past the end of the transition era and into the early 1970s. (Out of the 20 built by GE in 1949 for Russia but never delivered because of Cold War politics, three went to the South Shore and five to Brazil's Paulista Ry.) Part of this longevity was due to a late-1950s modification to the Little Joes to allow them to run in multiple-unit configuration with diesels.

The Milwaukee's famous bipolars were all out of service by 1960. By 1962, all were cut up except one that was saved and sent to the St. Louis Transportation Museum. Between 1970 and '71, the wire started to come down on the Coast Division. In June 1974, electric operations also came to a close on the Rocky Mountain Division.

A concern about modeling any railroad that runs through isolated mountain ranges is whether the gorgeous scenery is enough to offset the lack of heavy traffic, and closely spaced towns with interchanges with other railroads and industries to switch. I asked Bruce Carpenter, who models this part of the Milwaukee Road, about that.

13

Illinois Traction, also known as Illinois Terminal, hosted significant freight business through the transition era using Class B, C, and D freight motors. Class C 1597, top, has a short train well in hand as it crosses the Sangamon River north of Springfield, Ill. Upgraded Class D 70 in the later light-green scheme, left, rumbles through downtown Lincoln, Ill. Street running was a major liability for any electric railroad. The future in the form of a pair of Geeps stares 1592 in the face at East Belt Yard in Springfield, right. *Two photos: Sandy Goodrick, Kevin EuDaly collection; GP7 photo: Bernard Rossbach; all courtesy White River Productions*

"I shared your concerns about modeling the Milwaukee electrification through the Bitterroots. However, after speaking with former Milwaukee employees, the terminal in remote Avery, Idaho, became my main focus. It has proven to be very challenging for my operating crews. Avery was a tonnage-reduction yard, and trains set off or picked up tonnage based on power on hand and whether pushers were available. For the section of the Rocky Mountain Division I'm modeling, I was able to design the timetable to within 20 minutes of the actual timetable.

"Determining and calculating tonnages and power required to depart Avery has been fascinating to watch during sessions. Also, Avery is a crew-change point, something I've never seen modeled before. The dispatcher has to keep track of crew availability and expiration times, and if they need to be "deadheaded" to the east or west of Avery. It's either a ghost town, or there are two to four trains wrestling to get into town, get their work done, switch out electrics for diesels, and change crews."

Hauling copper ore in Montana was the raison d'etre for the Butte, Anaconda & Pacific. Its fleet of 17 boxcabs built by General Electric, **10**, was augmented in 1957 with a pair of GE road-switcher-like units that ironically would not operate in multiple with the older power. After a new ore concentrator was installed at Butte, the need to move ore to Anaconda was curtailed, allowing diesels to bump the electrics aside in 1967.

The Great Northern electrified its main line through the Cascades with several types of Baldwin-Westinghouse and General Electric boxcab electric locomotives. They also acquired handsome W-1 class streamlined electrics from GE. One of the Y-1 boxcabs, 5011, was extensively damaged in a fire and was rebuilt with EMD F unit cabs on both ends, creating an inviting kitbashing project for the GN modeler or freelancer, **11**. The electrification lasted until 1956 when improved ventilation in Cascade Tunnel allowed diesels to operate

14

A pair of aging Baldwin steeple-cabs made it through the transition era on the Chicago South Shore & South Bend. By the 1960s, however, rebuilt New York Central boxcabs and 800-series cab units had assumed most duties.

15

The Chicago North Shore & Milwaukee was best known for its commuter service but also hosted modest freight operations. Three steeple cabs move a freight at Rondout, Ill., top, in January 1961, and 459 (one of two ex-Oregon Electric motors) hauls tonnage through Lake Bluff, Ill., bottom, six month later. *Both: Bill Schaumburg*

Motors 1628 and 1622, built in the Pacific Electric shops, pull a 52-car freight westward over the Santa Fe tracks at Bench, Calif., near San Bernardino, on August 27, 1948. The SP boxcars are no. 15405 (Class B-50-15 with steel sides), 33689 (a 1937 AAR design), and an A-50-4 auto car. *Lewis Harris photo, Arnold Menke collection; courtesy Tony Thompson*

straight through from Skykomish to Wenatchee, Wash.

The smaller players

In the summer months, my parents often drove from our 1940s hometown of Sheffield, Iowa, north to Mason City ("River City" in native Meredith Wilson's "The Music Man"). We then turned west to Clear Lake, where friends had a lake cottage with a dock.

(In 1959, "the Day the Music Died," Buddy Holly, J.P. "Big Bopper" Richardson, and Ritchie Valens staged their last performance at the pavilion in Clear Lake. Their chartered Beechcraft Bonanza made it only a few miles before crashing into a field during a rapidly approaching snowstorm.)

That drive took us along the electrified main line of the Mason City & Clear Lake Railroad. I don't recall ever seeing anything move, but their "cab on a raft" freight motors caught my eye. It was hard to imagine that those things could muster enough power to move on their own, let alone pull anything.

Surprisingly, one can make the same drive today with an excellent chance of seeing several of successor Iowa Traction's ancient but spit-shined Baldwin freight motors hard at work moving loads of soybeans out of a large mill in downtown Mason City, **12**. Other Iowa interurbans hosted extensive freight operations that lasted in the 1950s. Also in the Hawkeye State, the Charles City Western, **8-7**, ended electrified service in 1968 following extensive damage caused by a tornado.

One of the most famous and extensive electrified interurban operations was the Illinois Terminal, later called Illinois Traction, which operated a Y-shaped system between Danville on the east, Decatur and Springfield in central Illinois, Peoria at the northwest corner, and St. Louis to the southwest. Its handsome arch-windowed interurban cars and class B through D freight motors, **13**, remain popular with traction modelers.

Diesels had gained a foothold in 1948 with the arrival of Alco S-2 and RS-1 units. EMD switchers were added in 1950 and '55, and GP7s joined the roster in 1953. Electrified freight operations ceased in 1956 and passenger operations in 1958. IT continued to operate portions of its main line as a diesel-powered freight railroad until it merged into the Norfolk & Western in May 1982.

During the transition era in Indiana, the Chicago South Shore & South Bend had similar Baldwin steeple-cab freight motors, **14**, some rebuilt former New York Central boxcab electrics, and three General Electric 2-D+D-2 streamlined cab units—what the Milwaukee Road dubbed Little Joes, as they were originally built as 5-foot-gauge locomotives for Joseph Stalin's Russia. They were retired in 1983; two survive in museums. Today, diesel-powered freight operations are under the South Shore Freight banner, but true interurban operations between the South Bend, Ind., airport and Chicago's Loop continue.

Insull-relatives Chicago North Shore & Milwaukee, **15**, and Chicago Aurora & Elgin (Insull lines eschewed commas in their corporate monikers) also operated electrified freight service during and beyond the transition era.

Passenger service on the 48-mile "Roarin' Elgin" abruptly and infamously ended in the middle of a day in the late 1950s; the railroad itself ceased operations in the early 1960s.

Pacific Electric had an extensive commuter operation that also accommodated freight traffic over portions of its Bay Area system, **16**. Tony Thompson notes that "PE built motors 1619-1631 in their Torrance shops, with GE electrical equipment.

Already well into the grade ascending the East Bay hills, Sacramento Northern motor 653 has a 16-car freight in hand in June 1947 as it crosses the Chabot Road bridge in Rockridge. Another motor is helping at the rear. Motor 653, delivered in 1928, was one of five standard GE freight motors. Number 653 survives today at the Orange Empire Railway Museum.
Arthur L. Lloyd Jr., courtesy Tony Thompson

Interestingly, they used the Baldwin steeple-cab body design, the only known case of a Baldwin-style body with GE equipment (Baldwin's electrical partner was always Westinghouse)." Numbers 1619-1626 were placed in service in 1924, the remainder in 1925. These motors normally operated in the northeast of the Los Angeles basin (PE's Northern District), pulling all kinds of freight in the San Bernardino area and towards the city. Electrified operations continued until 1963.

The 183-mile Western Pacific subsidiary Sacramento Northern, **17**, operated electric freight motors from Oakland north through the state capital to Chico, Calif., into 1965. Like most interurbans (all passenger service ceased in 1941), it operated on city streets in Oakland, Sacramento, Yuba City, and Woodland. The first diesels arrived in 1944, and it later became part of Union Pacific.

"There were lots of bits and pieces of the Sacramento Northern as it slowly went from mainline electric to isolated industrial spurs with GE 44-tonners over the Transition Era," recalls Bay Area native Jim Providenza.

Modeling footnotes

I won't devote a chapter to modeling electrified railroading, but I will offer a suggestion: Lionel once made a shortened E33 shell that fit onto a Geep chassis. It might provide the basis for an interesting project for freelancers who are building small layouts featuring electrified railroading. And it wouldn't raise eyebrows as much as, say, lettering a GG1 for a fictitious railroad.

Wiessmann makes a clever accessory for electrified railroads: a tiny blue-white LED that mounts atop a pantograph that flashes randomly to simulate arcing between the shoe and overhead wire. I saw it on an HO model of a Swiss railroad, and it was very convincing despite the overhead wire not being energized.

The takeaway for this chapter is that it's apparent that mixing steam and/or diesels with electrified freight operations, even classic passenger interurbans, isn't at all out of line during the transition era—as long as one makes plausible location choices.

CHAPTER FIVE

Those colorful branch and short lines

Among the most revered short lines was the Maryland & Pennsylvania. Modeling it once required the acquisition of a fleet of brass imports, a task greatly eased by Bachmann's introduction of several standard Baldwin steam locomotives. Stan White modeled a number of Ma & Pa scenes including famous Taylor's Trestle.

It's only natural to lament the passing and inaccessibility of what once was. Many of us can recall the late 1940s and the 1950s, but only a few have clear memories of the Second World War and even earlier times. So it follows that modeling the post-war era would be more popular than modeling a prior period. But there's a way to enjoy depicting the transition era and earlier eras at the same time: Model a short line, **1**. Such small-time outfits often subsisted on a diet of hand-me-downs from their big brothers. Even branch lines often saw the use of equipment that no longer met mainline needs, **2**.

Second-hand Roses

My freelanced Allegheny Midland layout connected with both prototype and freelanced railroads. AM trackage rights over the Western Maryland afforded an opportunity to run A-B-A consists of EMD Fs. (The AM dieselized too late to have purchased covered wagons.)

At its midpoint in Midland, W.Va., the AM connected with the Ridgeley & Midland County. This short line hauled coal and wood products out of the mountains east of the Cheat River using small steam power, **3**, and Alco RS-1s, **4**. A wood combine sufficed to take miners from town out to a coal mine and bring them home after their shifts were completed. There was no way a wood passenger car would have been used on the mainline AM of the 1950s in other than maintenance-of-way service. And the AM dieselized much too late for RS-1s to have joined the roster.

A similar tale played out on short lines across the continent. What the big guys no longer needed was just the thing to populate the roster of a less-than-flush short line, **5**.

Cutting costs

As railroads realized that the glut of riders typical of the war years was not going to be sustained into the 1950s, train-off-petitions became the norm. Railroads cut costs wherever they could, often in the form of gas-electrics, **6**, and then Budd Rail Diesel Cars, **7**, on both main and branch lines. *The Short Line Doodlebug: Galloping Geese and Other Railcritters* by Edmund Keilty (Interurban Press, 1988) provides myriad examples.

A few mixed trains, often with a coach or combine trailing the

2

The Chesapeake & Ohio ran a mixed train trailing a steel heavyweight combine up the branch from Covington to Hot Springs, Va., into the 1970s, left. The turntable at the end of the branch, right, was on the side of a hill mostly surrounded by open air. The combine may have been built for branchline use, but the Geep was originally mainline power.

freight consist, were still a staple of 1950s railroading. I recall riding the Chesapeake & Ohio's Hot Springs, Va., mixed in the 1970s with Jim Boyd, **2**. We also chased and rode the Norfolk & Western's legendary Abingdon Branch local, which in the '50s was powered by a 4-8-0 Mastodon and sported a steel combine. This is the train that O. Winston Link and Thomas H. Garver immortalized in *The Last Steam Railroad in America* (Harry N. Abrams, 1995). One of these odd beasts is now in regular service on the Strasburg Rail Road, **8**, giving freelancers an excuse to letter an LMB or Sunset brass import for their own railroad.

Several Southern states, notably Georgia, ran a number of mixed trains through the transition era, as Jim Boyd noted in the January 2017 issue of *Railfan & Railroad*. You may also find the photographs by John Krause and H. Reid in *Rails Through Dixie* (Golden West, 1965) inspirational.

The South was permeated with interesting short lines. Good examples are the original Norfolk Southern and the Durham & Southern, **9**, which intersected at Fuquay-Varina, N.C. Both hosted all-Baldwin rosters until EMD sold GP18s to the NS and GP38-2s to the D&S.

Narrow-minded survivors

My book *Guide to Narrow Gauge Modeling* (Kalmbach, 2014) provides an overview of narrow-gauge railroads, including those that survived into the transition era. Notable among those were the Colorado Rio Grande lines, which ran into the 1960s, the East Broad Top in Pennsylvania (1956), and the East Tennessee & Western North Carolina (1950), **10**.

The EBT is a good example of a railroad that survived not because of sentimentality or tourism but because of how it operated. It was primarily a bituminous coal hauler, with 22-car trains trundling north to a coal cleaning plant alongside the Pennsylvania's main line at Mt. Union in the central part of the Keystone State.

Since the coal had to be unloaded, cleaned, and sized before being

3

Small steam found continued employment on branch and short lines during the 1950s. Here Ridgeley & Midland County 4-6-0 no. 30, owned by Max Robin, brings bituminous coal into Midland, W.Va., for interchange with the Allegheny Midland.

4

The Rutland green and yellow paint scheme survived its 1960 abandonment on the Green Mountain Railroad's Alco RS-1s, top (also see 3-1). I followed this technique by re-lettering an HO Atlas Rutland RS-1 for my freelanced Ridgeley & Midland County, left, which interchanged with the Allegheny Midland at Midland, W.Va. Operating a shortline railroad in the 1950s allows continued employment of hand-me-down equipment from Class 1 railroads. This classic wood combine, right, scratchbuilt by Jim Boyd, took miners to and from work on the R&MC.

5

No two Minneapolis & St. Louis RS-1s had the same paint scheme. Number 546, left, modeled in HO by Clark Propst, wears one; Lake Erie, Franklin & Clarion's ex-M&StL 22, right, wears still another. The LEF&C unit serves as a guide to a freelanced model paint scheme: Just change the road name and maybe the number!

reloaded for transit to customers, there was no extra effort involved in unloading narrow-gauge twin and triple hoppers and reloading the prepared coal in standard gauge hoppers. EBT also avoiding breaking bulk—unloading a car of one gauge and reloading the lading into a car of the other gauge—by using an overhead crane to raise each end in turn of a standard gauge freight car and rolling three-foot-gauge trucks under the car for a trip south. Such cars couldn't negotiate any of the EBT's several tunnels, but they could get to the railroad's main terminal at Orbisonia/Rockhill Furnace and even head down the Shade Gap Branch.

The EBT also assembled a gas-electric, **11**, from parts supplied by Brill and used it to cut costs compared to steam-powered passenger runs.

The Denver & Rio Grande Western's line to Farmington, N.M., hung on by virtue of pipe-laying projects that sustained into the 1960s. In 1943, perennial favorite Rio Grande Southern happened to be in a position to secretly move what were formerly considered worthless mine tailings that contained uranium ore for the Manhattan Project. It then staggered along until shuttering its doors in 1951.

Logging lines

Mention logging railroads and geared locomotives—Shays, Heislers, and Climaxes—immediately come to mind. Their flexibility enabled them to negotiate crudely constructed and often temporary tracks that would see a conventional rod engine hit the ties before it had left the mill site. Most famous was California's narrow-gauge West Side Lumber Co. The WLC hauled logs that had to be unloaded and cut into lumber before being shipped to customers, so no added inefficiencies were encountered from forest to lumberyard.

The plethora of logging companies means that almost any mythical name has a ring of plausibility, a boon for the freelancer. And one can do first-hand homework even well into the 21st century, thanks to a number of tourist lines powered by geared locomotives. The most extensive example is Cass Scenic Railroad in eastern West Virginia, **12**, which boasts operating geared locomotives of all three major types. The line was operated by Mower Lumber Co. until the end of the transition era in July 1960.

6

The Minneapolis & St. Louis operated 11 gas-electrics built by St. Louis Car and Electro-Motive Corp. in 1929–'31. *Courtesy Gene Green*

7

Budd-built Rail Diesel Cars (RDCs) were gas-electrics' lightweight successor. They saw service on both branch lines and heavy-duty main lines, including the Baltimore & Ohio.

What had once been prime mainline power on the Norfolk & Western, 4-8-0 no. 475, wound up in a Pennsylvania gravel pit before being rescued by the Strasburg Rail Road, where it continues to see regular service today. Brass models of this Mastodon have been imported.

The Durham & Southern crossed the original Norfolk Southern at Fuquay-Varina, N.C., and both roads were Baldwin customers. The D&S units sported a classic Baldwin paint scheme: a wide band with rounded ends. Here a trio of Baldwin road switchers waits at Apex, N.C., with cars from the Seaboard Air Line.

The aptly named East Tennessee & Western North Carolina's narrow-gauge main line survived until 1950. One of their classic Baldwin Ten-Wheelers, no. 12—occasionally re-lettered for the ET&WNC—sees regular service at the Tweetsie amusement park near Blowing Rock, N.C.

East Broad Top gas-electric M-1 was a home-brew, assembled from parts supplied by J.G. Brill in Philadelphia. It is stored operational at Orbisonia/Rockhill Furnace, Pa., waiting for the day when the EBT resumes tourist operations. *Doug Leffler*

West Virginia's Cass Scenic Ry. boasts a varied fleet of geared locomotives ranging from Shays to Climaxes and Heislers. The 11-mile ride from the former C&O connection at the company town of Cass, once home to a large sawmill, up to Bald Knob is a visual and aural treat.

CHAPTER SIX

Modeling steam power

"The engine that saved a railroad" is how Nickel Plate Road authority John A. Rehor described the NKP's superb fleet of Berkshires in the October 1962 *Trains* magazine. Born in 1934, by the late 1940s the class S Alcos (and class S-1 Limas) were negotiating the formerly narrow-gauge line from Frankfort, Ind., to East St. Louis, Ill., without breaking a sweat save for a few notorious grades out of river valleys. Here NKP's first Berkshire, Alco-built no. 700, leans into a superelevated curve just west of Frankfort on my HO tribute to the NKP's St. Louis Division as it appeared in the fall of 1954.

You can choose an era to model or let an era choose you. Either way, there are major compromises to accommodate: Pick this, lose that. I'll cite some examples of choices that modelers made where their hands were forced by reality. As a result, one chose to freelance, the other to have a sizable fleet of 2-8-0s custom built. Despite today's plethora of superbly detailed, good-running steam locomotive models in every scale and gauge, difficult choices remain. But the rewards remain equally high, **1**.

Considering capabilities

By titling this chapter "Modeling steam power," as opposed to modeling a specific steam locomotive, the point is that we'll take a high-level overview of what one should consider when choosing to model any part of the steam era but especially the last decade and a half of steam locomotives. That the rewards exceed the liabilities is apparent from steam models' ongoing popularity with model manufacturers and importers and their customers. But, as with full-size steam power, there are liabilities we need to discuss.

One of the first surprises that typically greets the steam modeler is that the performance of a model locomotive may not equal its appearance. There's a comment in Linn Westcott's book about John Allen and his remarkable Gorre & Daphetid (pronounced "gory and defeated") HO railroad that meant little to me until I too faced the same problem.

Under a photo of a Pacific Fast Mail Chesapeake & Ohio 2-6-6-2, Linn noted that John never could get the locomotive to perform as well as he expected. I later acquired two of the same model, an early brass import with decent detail, and they didn't perform well for me, either. Until I had the suspension of the front engine rebuilt—it came from the factory with nothing more than a spring-loaded pin pushing down on a wide plate above the front six drivers ("engine")—it performed like a light 0-6-0. Afterward, each Mallet would haul about 12 loaded hoppers up a 2.5-percent grade and around 30"-radius curves.

Was that adequate? C&O documentation suggests the prototype would pull between 12 and 20 hoppers under similar conditions, so—after the rebuilding and expenditure of many hundreds of dollars (times 2)—all was well, **2**.

Three of my Key Imports brass Nickel Plate Road 2-8-4s were loaded with weight and would handle loaded 20-plus-car coal trains up those grades and around those curves with ease. When the Allegheny Midland was dismantled and they resumed service on the Nickel Plate Road, tackling the worst westbound grade, which was well under 2 percent, was and is no problem. I usually stage these "lead sleds" at the east end of the division so they are ready to handle westbounds up to the railroad's passing track length of 30 cars.

But other brass and "plastic" Berkshires were not up to the task. I still recall the sinking feeling I had when a test train out of Frankfort, Ind., stalled with a dozen cars. Alarmingly, efforts to add weight were unsuccessful on the mass-produced locomotives.

Bill Darnaby had encountered the same problem with his Maumee Route 4-8-2s, which began as Hallmark brass Illinois Central Mountains. After the initial panic attack, he tried changing freight-car wheel sets to free-rolling metal ones from InterMountain. That basically solved the problem for both of us. (It also facilitated Bill's subsequent installation of signals at interlocking plants, which required adding resistors to bridge the insulation on those wheelsets.)

Joe Borick of Cheat River Engineering rebuilt my two Pacific Fast Mail C&O H-6s so that the front engine actually contributed to the tractive effort. They subsequently performed very much like the prototypes, one of which—1309—has been restored to service on the Western Maryland Scenic Railroad.

In the 1960s, Indiana's Monon sported a varied roster of modern power, including no. 400, the first Century 628, brand new low-hood C-420 514, and high-hood 501 at Lafayette Yard in September 1967. Modeling this and earlier diesel power is appealing, but the resulting long trains greatly reduced train frequency.

4

Tender swaps are a relatively easy way to change the look of a stock locomotive. For example, the relatively short USRA tender of NKP no. 662, top, was stretched (no. 614, middle; note the six-wheel Buckeye truck under the coal bunker) or replaced (no. 639, bottom). Key Imports offered the stretched versions on some of their NKP Mikados, and the "Berkshire-style" tenders can be modeled with tenders from old HObbyline kits or AHM/Rivarossi 2-8-4s. *662: Lima; 614 and 639: Bill Raia collection*

Despite improvements that manufacturers and importers have effected since the 1970s, it still pays to check for these key attributes:

• Availability: Is the specific type of locomotive needed to fill out your roster still available?

• Affordability: Can you afford to buy enough of them to represent the prototype's fleet? Onesies and twosies are not convincing on any but the smallest railroad. You can argue that a single 2-6-6-2 in mine shifter or pusher service is the last of its kind and going to be retired soon, but you'll be fooling only yourself. (The timeworn adage that "It's my railroad and I can do whatever I want!" quickly implodes the moment you want to share the railroad with knowledgeable friends or book, magazine, or online readers.)

• Reliability: Will the engine perform well and continue to perform well on a regular basis? If not, can you correct its deficiencies or afford to pay someone else to do that? I outsource the installation of sound decoders and bringing the engine up to snuff before entering service. I now have steam locomotives painted and weathered, **2-12**, trading money for time I can better use to make progress on the railroad itself. And if a certain type of locomotive consistently balks when negotiating some curves and turnouts, be sure you can correct this. Crews will not forgive unreliable performers.

• Maintainability: Models, like their prototypes, wear out. Replacing parts is usually much easier and cheaper on a diesel than on a steam locomotive.

• Performance: Just because an engine looks powerful doesn't mean it will perform in a similar manner. If you're at the planning stage of your railroad, build a test section using the sharpest curve and steepest grade you are contemplating, and then run representatives of your locomotive fleet up and around the test track toting a desired-length train. It's better to be disappointed now than enraged later.

What's practical

Just because we really want to do something doesn't make it practical to do so. Possible? Perhaps. But let's be pragmatic.

Looking at what others have done isn't always a reliable guide. Their circumstances and objectives may differ markedly from yours. Moreover, a lot has changed—for the better—in the past decade or so, and what was impractical two decades ago is often within reach today.

I've shared this example before, but it's worth another take: As a few knowledgeable modelers, most of them professional railroaders or working in jobs related to the rail industry, tried their hand at dispatching their railroads using timetable and train-order rules, interest in "TT&TO" soon began to pick up. I first experienced it on the Midwest Railroad Modelers HO club layout in downtown Batavia, Ill., a groundbreaking model railroad that has sadly joined other fallen flags in the pages of history.

Several of today's most-respected modelers—Bill Darnaby, Dan Holbrook, and Bill Neale among them—honed their skills on that layout. Bill Darnaby, who grew up in central Indiana, was a Monon fan, and he built up quite a roster of first-

and second-generation Monon diesel power: EMD F3s and Alco Century 628s and 420s, for example, **3**.

By the time he and wife Mary Ann had their own home close to Bill's job at the Electro-Motive Division of General Motors, Bill knew what type of railroading he wanted to model. Alas, it no longer could include the Monon, the result of the unavailability of key steam models.

Why steam? Experience with timetable and train-order operations at Batavia had convinced him that was the most interesting and challenging form of train dispatching. Through his work at EMD, he became aware of what the diesel had done to win the battle of locomotives on the full-size railroads. A diesel's tremendous tractive effort and ability to be operated in multiple-unit consists meant that railroads could run much longer, hence fewer, freight trains.

But the need to run fewer trains was the antithesis of what one desires on a model railroad. So Bill needed to retire his diesel fleet and amass sufficient steam power to move his modeled period back into the 1940s just prior to Monon's dieselization.

Good luck with that! Even today acquiring a sizable fleet of Monon steam is challenging. And back in the Maumee Route's genesis period, modeling almost any railroad in steam other than the "usual suspects" was problematic.

The solution was to create a freelanced railroad that boasted the characteristics that made TT&TO operations fun, notably a single-track, dark (un-signaled) main line through the heartland of the Midwest where one could expect to cross a foreign railroad every ten miles or so. The Speedway of the Midwest, the Cleveland, Indianapolis, Cincinnati & St. Louis—the Maumee Route—was the result.

For Maumee's standard steam power, Bill chose the ubiquitous USRA light Mikado, **4** and **9-14**, and Illinois Central 4-8-2s from Hallmark. They had an easily fixed gearbox problem that kept the price down on the used-brass market. He gave them a new look with some detail changes and NKP Berkshire long-haul tenders from Rivarossi 2-8-4s.

If Bill were starting anew today, he has stated that he would probably model one of the New York Central's dark, single-track lines. That's because of the much greater availability, affordability, and reliability of the needed motive power and rolling stock.

Choosing the most advantageous time period to model is often a matter of finding out when a favorite locomotive class operated. Perry Squier chose 1923 when he modeled the Pittsburg, Shawmut & Northern, as many of the smaller engines such as this 2-6-0 and the passenger trains they pulled disappeared the following year.

Time matrix

A good way to see what opportunities, or pitfalls, lie ahead is to make a simple matrix with years across the top (about 1945 to 1960 for transition-era modelers) and a column of key features you'd like to model along the left side, **1-14**. I'll use my NKP layout as an example.

As a kid, I had the opportunity to watch the NKP's Berkshires and a Mikado or two at work on the St. Louis Division in the 1950s. One can't "model the Nickel Plate," as even a modest-size line like that is far too large to model as a whole. I picked the section that ran through my then-hometown: the (take a deep breath) Third Subdivision of the St. Louis Division of the Clover Leaf district of the New York, Chicago & St. Louis, better known as the Nickel Plate Road. And even then I could model only nine of the two-dozen towns listed in the employee's timetable.

All but one of those nine towns has an interchange with another railroad, which was the major criterion for choosing each town. Fortunately, I killed three birds with one stone by modeling my hometown, as it had an

It took three Alco RS-3s to deliver horsepower equivalent to one modern steam locomotive. But those three units occupy more inbound and ready-track space under a coal dock (see **7-3**), which can be a concern. When diesel consists must be made up or broken apart, I assign the task to a roundhouse foreman. *Bill Darnaby*

NYC steam modeler Tom Bailey suggests that a key to modeling a locomotive that is living out its waning years is to make the hot surfaces—smokebox and firebox—look warmer. Moreover, these are the first areas to rust. Tom's artist friend Bill Lewis showed him how to create this look, and he really likes the effect. He also prefers dust to be a buff color like crushed traction sand rather than gray.

interchange and one of few remaining water tanks in 1954.

But this brings up another aspect of modeling the late-steam era. The NKP had purchased its Berkshires with large tenders that held 22 tons of coal and 22,000 gallons of water. Both were sufficient to let those locomotives make non-stop trips over the entire subdivision. They then ordered some similar tenders to equip many of their USRA light Mikados, thus extending their range, **4**. These improvements allowed the NKP to retire the only coal dock on the subdivision and all but two water tanks. Had I absolutely, unequivocally wanted to model that wood coal dock, which was just east of my hometown, I would have forfeited modeling the Berkshire era, which began just before the coal dock was demolished, and with it other aesthetically appealing attributes.

I wanted Berkshires and Mikados with the "flying" number boards and Mars warning lights, **1**. I also wanted Alco PA-1s on passenger trains 9 and 10, and these "Bluebirds" shouldered Pacifics aside around 1948. Together, that set my earliest timeframe hack at about 1950.

And since I am a huge fan of Alco RS-3s, which were delivered in April 1954, **6** and **1-14**, that moved the back end of my time bookends up to early 1954. Steam was pulled off the St. Louis Division in July 1955 when the GP9s arrived, so that locked in the right end of my timeline. The last time steam exclusively powered the fall grain rush was therefore the autumn of 1954. Bingo! I had my modeling period well defined in short order.

Although he models 1923 rather than the transition era, Perry Squier's Pittsburg, Shawmut & Northern is another good example of using a time matrix to pick a year to model. As he explored the railroad's history, he discovered that some of the classic smaller locomotives such as 4-4-0s and the many passenger trains they pulled, **5**—keys to maintaining a challenging timetable and train order operating environment—largely vanished from the roster in 1924. He then began to focus on 1923 and found that most other key modeling and operational attributes existed close to that date. By stretching the canvas of plausibility by a few months here and there, not many anachronisms were required.

Servicing facilities

As I documented in *Space-saving Industries for Your Layout,* it pays to regard a locomotive servicing facility not only as a place to replenish consumables on locomotives but also as a local industry. Hoppers brimming with coal or tank cars filled with fuel oil arrive, as do gondolas or covered hoppers filled with fine sand for traction. Gons or hoppers filled with cinders from coal-fired locomotive fireboxes need to be hauled away.

Neither of the division-point engine terminals I modeled had a track devoted to spotting carloads of sand for the sand house. Instead, they used a self-powered crane to unload a gon from one of the engine-serving tracks. Thanks to a Walthers Burro crane and a decoder, **8-3,** this creates another task for a yard crew to perform during an operating session.

Remember my earlier comment about needing to understand the work rules that were negotiated between labor unions and railroad management? Such rules affect engine-terminal operations too. Road or yard crews picked up their locomotives near the coal dock as their tour of duty began and left them near the dock as it ended, often "on the pit" where a steam locomotive's complex rotating and oscillating machinery could be inspected.

Between the coal dock and roundhouse was the domain of the outside hostler. There was often an inside hostler who moved locomotives needing more attention or periodic servicing into the roundhouse.

I therefore employ a roundhouse foreman/hostler at both division-point engine terminals. His job is to service the locomotives between runs, which includes adding coal and water. How long does this take? He pushes one button marked "coal" and then another marked "water" to activate an ITT Products (formerly Miller Models) sound system. When the noise stops, that function has been completed.

The hostler has a list of trains expected to depart today, and what type of locomotive each is normally assigned. It's 1954, and two of the four divisions that radiate out of Frankfort are dieselized, and two are still in steam.

Two EMD GP7s or Alco RS-3s are normally sufficient for Toledo Division trains, but Peoria Division trains get three units, **6**. It's his job to consist or de-consist various units to have them ready by scheduled departure times, a simple task with an NCE ProCab. Similarly, he needs to have suitable steam locomotives (2-8-4s, 2-8-2s, or occasionally a 2-8-0 or 4-6-4) serviced, pointed in the right direction, and on the ready track for all St. Louis and Sandusky Division trains. If the general yardmaster and chief dispatcher agree that an extra or a section of a scheduled freight is needed, as it almost always the case several times each day, he has to find suitable power for that as well.

Signs of use, aging, and neglect

As the twilight of steam approached, signs of benign neglect became apparent. Locomotives that once exuded the pride of the railroad's mechanical forces began to show a patina of rust on cylinder jackets and smokeboxes. Shiny areas on grubby cabs and perhaps the tenders showed where someone had cleaned just enough of the locomotive to enable it to be identified.

NYC steam modeler Tom Bailey has long espoused showing signs of use and age. His kitbashed NYC H-5 Mikado makes it clear that its better days lie somewhere behind the tender, **7**. Tom uses a rust color to impart both evidence of a warm surface and signs of neglect. "Two other effects I add are road dust and side- and main-rod grease," Tom notes.

"Most modelers seem to favor a light gray for road dust, but I think it should be more buff than gray. The road dust comes from even well-maintained track, partly because engine crews sand the rail for traction, and the sand is crushed to powder.

"To prevent rust," he adds, "rather than polishing the rods, many railroads coated the rods and valve gear with a grease that was very brown (**8**). This helps us modelers avoid having bright nickel rods that are in sharp contrast with a weathered boiler. The grease picked up road dust along the way, so it was seldom of a consistent color."

After everything else is done, Tom uses an airbrush for road dirt and a very light dusting of black from above

8

Tom Bailey notes that rather than polishing the rods, many railroads coated them with dark brown grease to prevent rust. Tom usually weathers with oil colors, as on this NYC Hudson, using mostly combinations of earth tones: raw/burnt umber and sienna. "The siennas are a bit reddish to my eyes, good for fresh rust effects," Tom notes. "The umbers are more brownish, good for older iron oxides. I try to not mix them before applying but to vary them during application, so it looks natural, not painted."

to replicate the effects of soot. He has observed that steam locomotives dirty themselves within hours of returning to service, even after a heavy shopping.

"I watched Southern 2-8-2 no. 4501 coming into Chicago from trips in Indiana in the early 1970s," Tom recalls. "The locomotive had been cleaned before it departed Indianapolis, but when it got to Chicago, there were signs of slobbering around the smokebox door from what was probably a poor seal, the freshly silvered smokebox had a thin coating of soot on the top, and the smokestack was coated heavily with soot around the rim. Road dirt showed on the running gear and along the bottom of the tender. And that was probably after only 250 miles or so of fairly fast running.

"I've noticed the same effect with Union Pacific 4-8-4 no. 844 on numerous trips," Tom adds, "and the UP is fastidious about cleaning the locomotive before a day's outing. So the argument that 'my [shiny] model just came out of a shopping' is valid only if it has seen no revenue runs since shopping.

"Of course, during the waning days of steam, locomotives were seldom cleaned between runs unless going out on a fan trip. And even after a heavy shopping, often the only repainting/cleaning would be where the work was done, usually around the smokebox and cylinders, and sometimes around the air compressors and associated piping.

Tom adds one final comment about locomotive weathering: "To my eye, weathering makes a model locomotive look larger. Maybe that's because weathering highlights some of the details."

Weathering a steam locomotive properly requires some study of its operating locale. Areas with hard water will show white streaks running down from anywhere water and steam vapor escaped—whistle, pop (safety) valves, leaking joints, and so on. Throw in passage through a tunnel or two and a locomotive will look grimy in no time at all.

As the fortunes of the railroads waned in the late 1950s, maintenance was curtailed, which meant that ballast was not regularly refreshed on cleaned, which in turn resulted in mud being "pumped" upward as heavy locomotives passed by. Running gear reflected this condition.

Anything that got in the way

9

USRA light Mikado no. 557 is working the local at Woodlawn, Ill., on Jerry Hamsmith's Mount Vernon Division of the Chicago & Illinois Western. Frank Hodina detailed the Broadway Limited 2-8-2 with brass detail parts, a Precision Scale Delta trailing truck, and a long tender from a Proto 2000 2-10-2. The canvas cab awning and graphics help give it a Chicago & Illinois Midland-family appearance. *Jerry Hamsmith*

of general servicing and regular maintenance was discarded. This could include cylinder jacketing. Streamlining began to disappear during the war years except on a few appearance-conscious railroads like the Norfolk & Western and Southern Pacific, whose 4-8-2s and 4-8-4s continued to look like someone cared until the end of steam, **2-1**.

Speaking of aging if not neglect, you will find that you're realistically modeling another aspect of prototype steam: locomotives out of service for maintenance. One of my light Mikes almost pole-vaulted itself off the railroad when a screw securing the main rod to a driver disappeared. Only a fortuitously placed telegraph pole prevented it from taking a long dive to the floor.

The point is that if you are considering a railroad that solely depends on steam for motive power, you'll need to allow for vacancies in the roster due to failures. How long the disabled locomotives are out of service will depend on how adept you are at fixing them or the backlog of the repair shop you use, the complexity of the repair, and availability of spare parts.

Radii and easements

Prototypical appearance, especially of steam locomotives, is a function of radius. Our Lionel O-27 train sets proved that suitable compressed locomotives could negotiate ridiculously sharp curves, but we're discussing scale model railroading, not toy trains.

Fearing a loss of sales from the casual model train buyer, manufacturers continue to find innovative ways to make even a 4-8-8-4 Big Boy handle a rather sharp curve, and anything under a 24" radius in HO is certainly sharp.

The purpose of this book is to explore ways to model the late 1940s and the 1950s effectively, not to simply get by and plant the flag of victory by making utterly unrealistic compromises. That includes providing a physical plant (track and roadbed) that, while still sporting curves well below reasonable prototypical standards, at least allows them to look like they could survive there.

Bill Darnaby placed his 4-8-2s on curves of various radii to judge their appearance in terms of how far the cab hung out from the tender and the pilot swung out over the outer rail, both being deal breakers when carried to excess. He determined that anything much below 42" created an unrealistic overhang on either end of the locomotive.

This initially seemed excessive to me, as I had used 30" as the minimum radius on the Allegheny Midland, and I thought my fleet of Berkshires and 2-6-6-2 Mallets, **2**, looked reasonably realistic. But the AM was a twisting mountain railroad, not a Midwestern flatlands racetrack like the Maumee, which may have accounted for my eye being more accommodating.

As Frank Hodina and I designed the Nickel Plate layout that currently graces my basement—the same space that had contained the AM—I ran the same tests Bill had and, to no one's surprise, came to the same conclusion: A 42" radius was the goal.

The real surprise: It was no more difficult to fit curves of the much larger radius in the same basic footprint that the AM's 30" curves had occupied! The lesson is to be careful what you decide to aim at, as you're likely to hit it.

I should add that on the AM, Maumee, and NKP layouts, easements make the transition between the tangent (straight) track and the constant-radius curves. My tangents are drawn using a bent wooden yardstick to connect the tangent with the 42" curves, which were drawn with about a ¾" offset (that is, the center of the curve is located 42¾" back from the tangent line that approaches the curve). Easement length is longer than the longest car (a full-length passenger car).

I also superelevated curves starting where the easement begins by raising the outer ends of ties using 1⁄16"-thick basswood shims. I sanded the shims starting at roadbed height so they sloped very gently upward to the full height of the shims in about two feet. I did the same on the Allegheny Midland's sharper curves and never experienced any derailment problems.

Freelancing power choices

For those who want to freelance in a plausible manner, the USRA series of locomotives is often an excellent choice. For starters, they're available from a number of sources in most popular scales from Z to large.

By far the best choice, Mikado authority Ray Breyer points out, is the ubiquitous USRA light Mikado (2-8-2). "Far fewer roads had the heavy Mikes," he points out, "and most of those went to the Milwaukee."

Good examples are the Mikados Frank Hodina detailed and painted

A dock switcher backs up the approach to get behind another cut of ore arriving behind a powerful 0-10-2 for the ore docks at Duluth, Minn., on Jeff Otto's proto-freelanced Missabe Northern Railway. It's closely based on the Duluth, Missabe & Iron Range, but some freelancing added modeling latitude. *Jeff Otto*

for the freelanced Chicago & Illinois Western, a railroad closely based on the Chicago & Illinois Midland, **9**. Both Frank and Jerry Hamsmith model portions of the C&IW (see *Model Railroad Planning* 2008).

The USRA 0-6-0 and 0-8-0 switchers, **2-11**, are other excellent choices for switchers on a freelanced steam-era railroad. They're available in most scales.

Sometimes a generic look is not what's needed, however. Jeff Otto models a prototype-based but freelanced iron ore railroad, which gives him enough latitude to work with commercial offerings. But from the outset he knew that his HO scale Missabe Northern, **10**, had to closely resemble a well-known ore hauler, and he selected the Duluth, Missabe & Iron Range as the base prototype. (See *Model Railroad Planning* 2013 for more details and photos.)

Smoke effects

So far, no one has invented black smoke that is harmless to breathe. The white smoke that some model locomotives can generate does create appealing visual effects when shooting videos or entertaining casual visitors.

However, the haze and perfumey odor would quickly be a bit much for crews to cope with during an operating session if many of my steam locomotives were so equipped.

The bottom line

A flawlessly performing steam fleet of any size is a commendable goal but rarely achieved during a several-hours-long operating session. Engines that ran flawlessly as you staged the railroad will assert their independence in remarkably creative ways as the fast-clock hands begin their orbits.

I usually have enough spare locomotives tucked away in the roundhouse to replace any recalcitrant beasts that fail out on the road or in the yard. If things have been going well, the replacement power may be run as an extra out to the site of the disembowelment; if not, I simply carry the new locomotive to wherever it's needed.

Such events occur with alarming frequency as new locomotives join the roster. Over time, the problems are usually sorted out and headed off at the pass. But you can never assume that a locomotive, once cured, will never ail again. That may lead to thoughts of following the prototype's transition era example and retiring steam in favor of more diesels.

I've had those thoughts. But then I watch—and hear!—one of those big Berkshires or handsome USRA light Mikados race through the croplands and lean into a superelevated curve, and all is forgiven.

CHAPTER SEVEN

Modeling early diesel power

Cliff Powers models the New Orleans Union Passenger Terminal, which was served by most of the major railroads in the South—Illinois Central, Kansas City Southern, Louisville & Nashville, Missouri Pacific, Southern, Southern Pacific, and Texas & Pacific. It would be hard to find a more colorful gathering of diesel motive power—here a bevy of elegant E units—than NOUPT in the Transition Era. Cliff documented his efforts to model the terminal right after it had opened in 1954 in *Model Railroad Planning* 2014. *Cliff Powers*

Modeling the diesel component of the transition era is almost too easy. Manufacturers and importers have blessed us with not only a wide variety of diesel models, some rather obscure, but also with railroad-specific detailing, **1**. In many cases, it would be hard for even a skilled modeler to do a better superdetailing job than the one done at the factory. But to focus so tightly on details is to ignore the bigger picture of how diesels can support our efforts to create models depicting this important era.

A pair of Alco FA-1s on a hot reefer train are more in keeping with the overall proportions of the train than adding a B unit or two would have achieved on Allen McClelland's original Virginian & Ohio HO railroad. *W. Allen McClelland*

Utilizing the victor's skill sets

As was the case in our Chapter 6 overview of the implications of choosing to model steam power, in this chapter we'll look at the pros and cons of opting to put more emphasis on a fleet of internal-combustion locomotives.

Many of the advantages accruing to prototype diesels also favor models. Let's review a few of the ways that the odds are stacked in favor of dieselization in any scale:

All driving wheels are powered, and all of the locomotive's weight is on them. Perhaps I therefore shouldn't have been surprised when one of my Proto 2000 2-8-4s spun its drivers with a dozen cars on a 1 percent grade. On a hunch, I replaced that steam engine with a single Proto 2000 Alco PA-1, and the diesel walked away with that train as though it were not doing any work whatsoever. I must admit that the model PA was "cheating" in that, unlike its prototype, all six axles were powered. But still …

I remember Allen McClelland telling me he discovered that one diesel would pull a lot of cars up the climb from Dawson Spring to Sandy Summit, Va., on his original Virginian & Ohio. But one unit up front looked a little silly, so as I recall he powered only one truck on some units, thus reducing tractive effort while maintaining good electrical contact. He also added dummy units to fill out the consist.

With diesels, the need to build 3- or 4-unit consists to equal the power of a modern steam locomotive means that a typical diesel consist occupies more ready-track space. This may reduce train length by a car or two if passing siding length is a constraining factor, and they can cause crowding in the engine terminal. *Bill Darnaby*

Allen was careful not to overpower a train in a visual sense too. Many trains had a pair of units up front, **2**, rather than an A-B-B-A set of Fs or FAs that might have been a quarter as long as the train they were pulling.

Compared to a steam locomotive, the drive mechanisms are simpler. On some model steam locomotives, only one pair of drivers is powered by the motor-driven gearbox; the side rods transmit power to the other driver pairs. Since the drivers are doing real work, this means that there is likely to be considerable wear on the bearing surfaces of the side rods.

The consequences of a part failing on a diesel may cause the locomotive to come to a screeching halt, but it will seldom derail dramatically as did a light Mikado after having the main rod come loose and pole-vaulting itself off the railroad, as I described in Chapter 6. That could have been a very expensive excursion off the high iron.

Details added to a diesel body shell can be delicate, but they're usually not nearly as fragile and prone to damage as the myriad pipes and appliances attached to a steam locomotive. You can make it very clear to your operating crews that in case of a derailment of a steam locomotive, you and only you are allowed to re-rail it. Good luck with that. A person's instant reaction is to fix the problem—hey, it's only the pilot truck that derailed. Besides, the owner isn't anywhere near the scene of the crime at the moment. And we all assume that somehow this is our fault

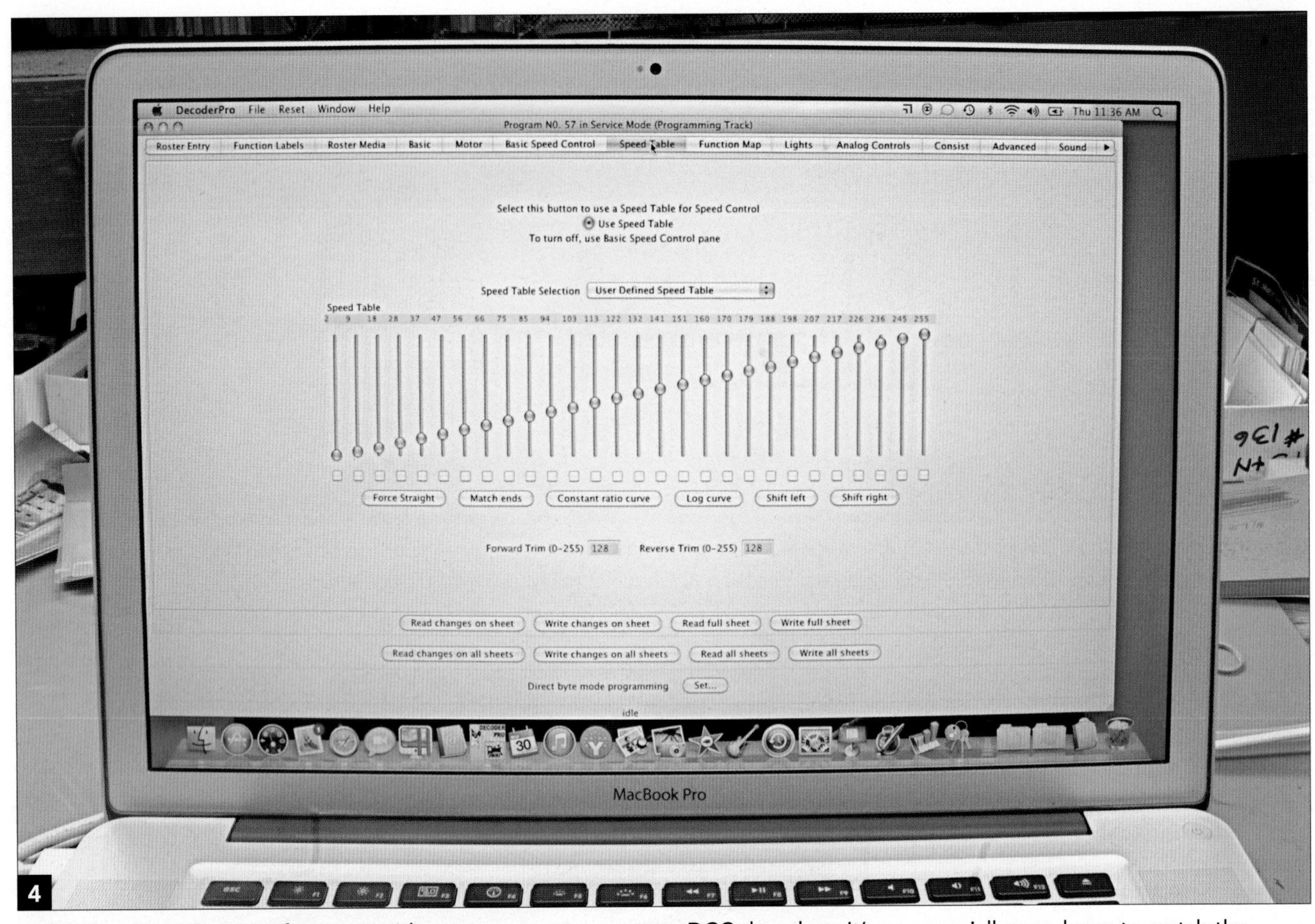

4

JMRI's free DecoderPro software provides an easy way to program DCC decoders. It's an especially good way to match the speed characteristics of diesel locomotives that will be arbitrarily assembled into consists. Here Perry Squier has set a decoder's speed steps for linear acceleration for one of his 2-8-0s.

5

This freelanced Tennessee Southern paint scheme by Allen Whittaker evokes the typical EMD styling elements that make a mythical railroad's livery fit plausibly into the transition era. *Allen Whittaker*

and don't want anyone else to know about it.

You'll do the most damage to your steam or diesel models moving them from workbench to layout or restaging the railroad. You therefore need to decide early on whether you're building a museum diorama with movable parts or constructing an operating railroad.

And if that railroad requires more than one person to operate, you need to take a deep breath and smile whenever someone inadvertently damages a model—assuming you want him or her to show up at a future operating session. If a crewmember proves to be a real klutz, you have a personnel problem to deal with. For the layout owner, an operating session is often more like managing (unpaid) direct reports than it is playing with trains.

Changing of the guard

Modeling the transition era means that you'll have a mix of steam and diesel power on your roster as diesel units begin to replace steam locomotives. But this is not a one-for-one swap.

The graph of the performance of a Pere Marquette Berkshire (**2-7**) shows that it topped 5,000 hp. Even three first-generation diesel units couldn't equal that. But those diesels could easily start the same train. Indeed, former steam engineers (meaning all engineers in the transition era) who were used to backing up their trains to "take slack" and thus start only one car moving at a time quickly learned that this was not only unnecessary with diesels but also likely to break a knuckle or cause a drawbar to fail if they really "leaned into it." But the diesels couldn't pull the same train as fast; horsepower is related to speed. Once a modern steam locomotive got its train rolling, it was off to the races as its output climbed the horsepower curve.

Of course, our model diesels will do almost anything we ask of them in terms of tractive effort and speed. The problem is their length: Three or four units in consist take up a lot of space on the engine terminal's in- and outbound tracks, **3**, on arrival and departure tracks in the yard, in passing sidings where they reduce the capacity by at least a car length, and in the roundhouse, where consists have to be broken up before units can be turned or stored in a stall.

Cost also enters the equation, as four diesels will cost more than one steam locomotive to acquire and to equip with DCC sound decoders.

And do some testing if you plan to use six-axle units like EMD SD7s, SD9s, and E units and Alco RSDs and PAs, especially if you superelevate closely spaced reverse curves. I had to do some tuning to my Proto 2000 PAs to make them comfortable negotiating the sinuous climb out of the Wabash River Valley, even though my minimum curve radius is 42", all curves have easements, and I limit superelevation to 1⁄16". Some brass RS-3s had unexpectedly stiff chassis that required rework at the importer to allow the trucks to "rock" as they negotiated reverse curves.

6

EMD's graphic designers found a clever way to ensure that the 5½" stripes applied to the sides of the Nickel Plate's Geeps, top left, retained that width when turning the corner and angling up the nose. This bit of geometry eluded the painters at Alco, top right, and the angled stripes wound up somewhat narrower. (Note the poling pocket near the "F" on 569's frame; see Chapter 9.) Bill Darnaby used the NKP livery as a basis for the freelanced scheme on the Maumee Route. *Maumee: Bill Darnaby*

The value of sound effects

Speaking of sound, we'll explore the use of sound for operational purposes in Chapter 9. But it's far more than just a novelty.

I ran some tests to see how long it took for an engineer to pick up a locomotive on the ready track, move it down to the departure track, couple onto the train, and "leave town." I did this with the DCC sound decoder silenced and with it operating normally.

The difference in time consumed was astounding, something like three to one during some phases of the departure sequence. When the locomotive was making realistic sounds, the crew had a means to gauge its speed and enjoyed hearing it go through its paces. Without sound, there was no aural feedback at all and each move tended to be made faster.

This was especially noticeable as the train eased out of the departure track. Cranking open the throttle before the caboose cleared the yard was instantly noticeable. Peer pressure works!

First-generation diesels also transitioned from having all motors in series to series-parallel and then to parallel as speed built up, as we have previously discussed. Most sound decoders that replicate first-generation prime movers (EMD 567s and Alco 244s and 251s, for example) simulate the transition process. Hearing the prime mover rev up and pull at a high rpm as the train starts, drop to idle, rev again, hold that for a time, drop to idle once more, and then resume high revolutions is an aural treat. And it leads an engine crew to open the throttle slowly so as to maximize these audio effects.

7

Wabash control of the Ann Arbor provides a sterling example of how a factory-painted model can easily be modified for a freelanced railroad with new road name and herald decals. Ann Arbor Alco FA-2 56, above, heads a train three miles north of Ann Arbor in 1947 wearing a scheme based on parent Wabash (top).
Wabash: Tony Koester; AA: Robert A Hadley

The volume or intensity of sound is a controversial topic. Most model railroaders are of an age where some hearing loss, maybe a lot of hearing loss, is to be expected. So what you hear may be somewhat different than what I hear. And no matter where we set the decoder volume, a busy engine terminal with a half-dozen or more diesel locomotives idling away is going to be noisy. But a dead-silent engine terminal is a like a cemetery.

I look at it this way: Railroading isn't a quiet activity. When one of those big Berkshires leaves town, I want everyone to notice. When an Alco PA cranks it up as it leaves the depot, that should be a memorable event for both visual and sound reasons. And when that Alco S-4 yard switcher hustles down the lead, I want the turbocharger's whine to catch your attention.

There's also a dramatic difference between sound levels that you set when you're alone with the railroad and those that are audible during a busy operating session. The noise created by conversations among crews will drown out much of what you perceived as an adequate level of sound from a decoder. Don't assume that workbench settings are loud enough.

The transition era featured a cornucopia of locomotive sounds: the hollow "Whuff!" of saturated steam engines vs. the crack of the majority equipped with superheaters; EMD vs. Alco vs. Baldwin vs. Fairbanks-Morse engines of various vintage; horns that ranged from the almost rude blat

8

You can see how a familiar paint scheme might look on a freelanced model if one of the colors were changed by using photo-editing software. Here Erie Lackawanna F7's maroon and a Bessemer & Lake Erie SD's orange were shifted to blue. Note the F's point-tipped stripe, a very typical EMD design element.

Jack Ozanich aptly chose a variant of a standard EMD striping pattern for his freelanced Atlantic Great Eastern. Similar schemes could be found nearby on the Maine Central, Boston & Maine, and Lehigh Valley, lending the AGE the desired New England flavor. *Craig Wilson*

of a single-bell "duck" horn to the melodious chime of a multi-bell horn or Hancock air whistle; bells that rang in syncopated time with the fireman pulling on a lanyard vs. rhythmic air-operated bells that had none of that school's-out charm. If our decoders miss any of those nuances, our modeling suffers accordingly.

Background sounds also play role. I spent some summers on my grandparents' farm in southern Iowa. They had a John Deere B model and a Farmall M. The JD's two-cylinder engine gave them the nickname "Poppin' Johnnies," and to this day I can't pass by one of those green and yellow machines without paying my respects. Thanks to Bill and Mary Miller, who used to produce sound modules (they're now available from ITT Products), I even have a sound module that plays the distinctive cacophony of a John Deere two-lunger firing up and going about its business.

I also have two of their birdsong sound modules, which provide a nice sunny-fall-day ambiance to the railroad room. (At least one of my crew hears this as irritating noise!)

Speed-matching decoders

If you double- or triple-head steam locomotives, you should ideally have one crew for each locomotive. That means the locomotives need not operate at the same speed at a specific throttle setting, as the engineers can adjust the speed to avoid wheel slip or sliding.

Since diesels are usually consisted, it is more important to adjust the decoders so that they all operate at the same speed for a specific throttle setting. Using Java Model Railroad Interface's free DecoderPro software is by far the easiest way to do this (jmri.sourceforge.net), **4**.

As railroads were getting used to operating their new diesel locomotives, however, they tended to keep like units in consist. Operating consists comprising only one model manufacturer's cab or hood units may preclude the need for speed matching, but as the 1950s matured, mixing locomotive types was commonplace. The first consists I saw on the NKP's St. Louis line after it was dieselized in July 1955 were a mix of GP7s, GP9s, and RS-3s.

This was not universally the case, however. Some manufacturers (notably Baldwin) employed air rather than electrical controls when in consist. Be sure your consists reflect the characteristics of the units you model.

Speaking of consisting, if that's an important aspect of your operations, as it is on my railroad, be sure the DCC system you're planning to buy makes consisting very easy and flexible. I use NCE's system, for example, and consisting is a snap for my roundhouse foremen: They select the lead unit's road number (address), enter forward or reverse, the trailing unit's number and F or R, and then any middle units F or R. When a consist is established, there's no need to remember or record the consist number; just key in the lead unit's road number. And either the

10

Mike Confalone's Allagash Railroad is set in Maine. Mike has moved the modeling time period past the transition era, but a lot of first-generation power continues to see service. It interchanges with Jack Ozanich's Atlantic Great Eastern, adding interest to both railroads' operations. *Mike Confalone*

lead or trailing unit can become the lead unit simply by selecting its road number. There is therefore no need to run a consist in reverse as it works its way back to the originating terminal.

Prototype vs. freelancing

Prototype modelers have a lot of the problems that cause freelancers to scratch their heads already resolved for them. They know in advance what colors the railroads they've elected to model painted their lineside structures, for example. They know precisely what type of cabooses operated on each part of their railroads and how they were painted and lettered during the period their railroads depict. And they know what type of diesel freight, passenger, and yard units were on the roster, in what quantities, and how they were painted.

For the freelancer, however, the sky's the limit in terms of diesel roster choices and paint schemes. Or is it? Actually, the freelancer has a lot of homework to do if plausibility is a consideration. As we briefly discussed in Chapter 3, that's because only a handful of graphic designers, most of whom worked at Electro-Motive, created the paint schemes worn by first-generation diesels. It follows that any freelanced paint scheme of that period should studiously follow the prototype railroads', and often as not EMD's, lead, **5**.

Put another way, if your freelanced railroad began to dieselize with, say, Baldwins, the paint scheme will be markedly different than a railroad that dieselized with EMDs. And units subsequently acquired from a different builder will logically sport a paint scheme based on that of the original units, albeit slightly modified to reflect the secondary manufacturer's interpretation of the graphic design, **6**.

It's not difficult to create a convincing depiction of a transition era railroad even if you're freelancing. One of the easiest ways to do that is to select a handsome factory-painted model of a scheme you like, one that looks at home in the region of North American you're modeling, and simply re-letter it. The prototype did it, after all, **7**!

Taking this a step farther, you could change one of a commercial model's colors, painting over Erie Lackawanna's maroon to blue, for example, **8**. You can use photo-editing software to change the colors before you take the plunge. Using Photoshop Elements, for example, just select the image of the unit whose paint color you'd like to experiment with, then choose Enhance > Adjust Color > Replace Color. Set the "eyedropper" on the section of color you want to replace and then play with various colors until you find one or more you like.

Or you could buy a commercial decal set with the needed striping, apply the base color(s) of your choice,

Locomotives types tended to develop their own distinctive weathering patterns. EMD Fs and Es often generated a "bow wave" as grime and mud from the roadbed sprayed up each side behind the pilot. North Western F7 415 shows this pattern. *J.F. Hilton*

and substitute your own road-name decals.

But right up front, you need to realize that freelancers actually have more homework to do than prototype modelers. Prototype modeling is more a task of editing than creating; you do your best to find out what your chosen prototype did for a living, select a manageable segment thereof in terms of both geographical location and era depicted, and try to model that accurately. Anyone familiar with that prototype can instantly make a value judgment as to how well you've accomplished your goal.

Can't narrow your choices to a specific season and year to model and therefore plan to model "the 1950s"? Remember retired *Model Railroader* managing editor Jim Kelly's sage advice: "Saying you're modeling the 1950s means you're really modeling 1959 but doing a lousy job of it." Paring down your choices is not only a key step on the road to plausibility but is also less expensive. You'll learn not to buy anything outside of your chosen time parameters just because it's an appealing model.

The freelancer who aspires to depict prototype railroading of the 1940s and '50s has to be very careful not to stray too far from the norm. He's already walking on soft turf in that his "prototype" doesn't exist. It won't take much of a deviation from the familiar to derail his aspirations. His visitors, be they in person or via print or digital media, are likely to know what the real environment looks and feels like. Anything the freelancer does that is a bit too inventive will stand out like a gold-painted steam locomotive. (There was at least one short-lived example of that, a Burlington 4-8-4—and it did indeed stand out!)

That is not to say that prototype-based freelancing is overly difficult or going out of style. Consider such highly regarded examples as Allen McClelland's Virginian & Ohio, Bill Darnaby's Maumee Route, Jack Ozanich's Atlantic Great Eastern, **9**, Jeff Otto's Missabe Northern, **6-14**, and Mike Confalone's Allagash, **10**. And even many modelers whose layouts depict a specific prototype railroad are modeling a fictitious branch or division.

Prototype modeling is increasingly popular mainly because it's much easier to do that today than it was a decade or two ago. Model manufacturers and importers have gone the extra mile in offering exquisitely detailed models for a range of prototypes. As this is being written, for example, Bowser has just announced a new series of late-model (Phase 3) Alco RS-3s that have railroad-specific details that in the past have been difficult to create and apply. Among those they've announced are the exact RS-3s I need, but it wasn't even on my list of things I dared hope would happen in plastic any time soon. (Overland had done them in brass.) Patience continues to be a virtue.

Is this a bad thing? Is having super-detailed plastic Phase 3 RS-3s in both the NKP's as-delivered and post-1959 paint schemes a case of having models handed to me on a silver platter that I should have detailed and painted myself? Is it an example of the hobby shifting from model building to ready-to-run? Are we collectively losing our interest in model building?

I see it as a shift toward realistic operation as a primary goal. I have enough other projects to attend to on

12

A new diesel shop that supplements rather than replaces a roundhouse, as Cliff Powers built for his New Orleans-based HO railroad, will require several extra square feet of layout space and therefore is an essential part of early layout design efforts. *Cliff Powers*

my railroad and hence truly appreciate such support from the model railroad industry. I don't think this diminishes my standing as a scale model railroader one iota.

And consider why I am building a segment of the NKP's St. Louis Division in the fall of 1954. It's not to entertain myself with building models, although I do enjoy almost all aspects of model building as time permits. The goal was to build a time machine that would transport me and my crew back to some memorable days of my youth. I remember what I saw at lineside in the last four years of steam operation on that division, but I then understood what was happening imperfectly. And I saw only a tiny segment of the action on that line, as I wasn't old enough to drive or informed enough to ask my dad to take me to other locations.

By building a model railroad that depicts the railroad and its environment to an acceptable degree, I can now experience what I missed. It's not quite as dynamic, but it will do. To get there from here, however, I had to—and will continue to have to—take shortcuts. I kitbash structures, or I use almost stock kits as stand-ins for structures I still don't know much about. And I outsource some tasks. If a model manufacturer is gracious enough to supply precisely the models I need in ready-to-use form, I very much appreciate that support.

Moreover, I believe that until the prototype modeler realizes and acknowledges he or she is freelancing to a considerable degree, progress will suffer. We can't know everything we want to know, at least in the short term. If we sit on our hands until we do, the railroad-to-be will continue to be a good idea unrealized.

And if we can buy products that are reasonable models of needed prototypes, it behooves us to acquire them. If we don't, the manufacturers can make one of two choices: They can decide their model wasn't good enough for whatever reason and try harder next time; or they can decide that making models for that prototype railroad isn't a good economic bet. Many of them will make the easy choice: It's apparently not a popular railroad to model and hence not worth the effort for them to manufacture models for that prototype. There's no reason to buy junk, but we should consider rewarding a good effort if it eases our modeling pains.

Weathering

Just as steam locomotives tend to weather in predictable ways, so too do various types of diesels have distinctive weathering patterns. Consider carbody units such as EMD Es and Fs and Alco PAs and FAs. Notice the "bow wave" of dirt that sweeps up from behind the pilot, **11**.

Alcos were turbocharged, and the lag between fuel injection and turbo spin-up often caused plumes of oily black smoke to boil out of the stack. As expected, the roof on all sides of the single exhaust stack was coated with black residue, and it often ran down the sides of the unit.

When Tom Bailey and I were chatting about weathering steam locomotives (Chapter 6), Tom noted that it's impossible to make a steam locomotive look right by using only one weathering color. It's the mix of hues from black to rust that create a convincing appearance.

13 Alco's A1A (unpowered center axle) trucks spread the weight of the RSC-2 and -3 road switchers on light branchline rail. Note the sideframes with solid-bearing journal boxes, which do not resemble the ones used on C-C units. *Trains magazine collection*

Joint serving facilities

If the railroad you're modeling simply adapted its steam facilities to accommodate diesels, adding fuel and watering cranes for the diesels and unloading cranes for tank cars, servicing them in the roundhouse, and filling their sand boxes from piping already on a coal dock, not much additional space will be consumed.

But if they built a brand new diesel shop, **12**, that can consume several square feet on a layout. The building and related sand towers and fuel cranes make eye-catching models, but space must be allocated.

One way to save space is by building only half of the diesel shops up against a front-surfaced mirror. Or one side of the shops can be built abutting the fascia with the structure cut away to reveal the interior. This will make it easier to know which units are inside being serviced.

Special-service units

We've already discussed EMD's ill-advised attempt to make a special branchline unit dubbed the BL2 in Chapter 3. But there were units that were more adept at some tasks than others.

Milwaukee Road, Seaboard Air Line, Soo Line, and Union Pacific, for example, ordered Alco RSC-2 and RSC-3 road switchers, **13**. These are RS-2s and RS-3s that rode on A1A trucks—six-wheel trucks with un-powered center axles. The extra tractive effort of two more motor traction motors wasn't needed, and the extra axle per truck spread the weight out over the light rail typical of branch lines.

They were produced from 1946 into 1952 and are easy to spot: The axles are evenly spaced, whereas the center axle on RSDs is closer to the pilot end of the unit. Kato produced an RSC-2 in N scale.

The SD designation of EMD's SD series stood for "special duty," and they were initially seen as more suitable for use on heavy drags such as loaded coal trains. The NKP bought both SD9s and RSD-12s, for example, and they were assigned to the Wheeling District to lug coal out of southern Ohio. Today, six-wheel trucks are the norm, although the SD designation remains.

A 1937 agreement to protect firemen's jobs required two crewmen on locomotives weighing more than 45 tons. That accounts for General Electric's popular 44-tonners, produced from 1944 to 1956, **14**. Many Class 1 railroads had a few of them for jobs not requiring the "fireman." They and the later 70-tonners (1946-1958) were popular with short lines.

Thanks to Bachmann and others, modeling these diminutive but capable units in most scales is not a problem.

14 GE and other small-locomotive makers introduced 44-ton center-cab locomotives in response to a 45-ton rule that allowed operation without a fireman. Major railroads often bought one, as was the case with the NKP's "Mighty Ninety," or just a few for specialized assignments. Many short lines opted for the heavier 70-ton end-cab unit. *NKP 90: Jim Boyd*

CHAPTER EIGHT

Modeling the physical plant

It's the early 1950s in Cayuga, Ind., where the Nickel Plate crossed the Chicago & Eastern Illinois. The photo shows a watchman's shanty, to be sure, but there's so much more "behind the scenes." This is a scan of a drugstore copy of a print from a negative. This small black-and-white print revealed information about a two-tone paint job on the joint NKP-C&EI tower (apparently two-tone green, the C&EI Historical Society reports), a split-rail derail and piping back to the tower, several shanties, a bumping post, a crossbuck, and a corncob burning pit—and that's just the first layer of information! *Charlotte Schwab Miller collection*

The transition era was the last chance to get a good look at railroading as it had been from the industry's formative years. Although often somewhat worse for wear, all of the accoutrements of decades past were still front and center. Depots, towers, water tanks, coal docks, roundhouses—the stuff that modeling dreams are made of—still existed. Fortunately, so did good cameras, along with railfans and modelers who were growing increasingly aware that now—right now!—was the last time to document these vestiges before modernization took its toll.

Photos are our friends

Modeling a specific time period earlier than, say, yesterday is going to require that we acquire and refine our photo interpretation skills. We're not always lucky enough to discover that some kind soul had us in mind 50 years ago when he or she walked around a building and took sharp photos of it from every angle.

And often as not it isn't the main or intended subject of a period photo that is of the most value. A snapshot of the Nickel Plate Road's Division St. crossing watchman's shanty in Cayuga, Ind., **1**, is a good example of using a photo to see what else is evident in that image.

In this case, it's one of the only photos of that shanty I have ever seen. Charlotte Schwab Miller, whose father was the day-trick operator at the Nickel Plate Road-Chicago & Eastern Illinois interlocking tower, seen in the background, saved this photo and several others that have proven critical to my modeling needs.

What can we see? For starters, using precisely spaced wood or styrene board-and-batten sheet stock isn't going to do a very good job of capturing the true texture of the shanty! Behind it is a bin for the coal that heated it during the winter months.

To the left of the shanty is a pair of interlocking rods from the tower. They end abruptly at a split derail, which tells me that this was located just west of Division Street. I remembered seeing one to the west of the C&EI diamonds; now I know where the westbound derail was located. And that pole to the left of the shanty announces "End Automatic Block"; now I know where it was too and about how high the sign was mounted.

Squinting a bit, I can see the type of bumping post on the end of the house track that ends at the tower, and to the left (south), I can clearly see the style of crossbuck that the C&EI used to protect crossings on its Chicago-to-Evansville, Ind., double-track main line. I can also see that the NKP used three arms on its telegraph poles east of the tower but apparently only two to the west.

2

The Maine two-footers didn't survive into the post-war era, but they came close and remain an intriguing option for narrow-minded modelers such as John Rogers. This is Strong, Maine, where the Y-shaped Sandy River & Rangeley Lakes split into two main lines. Following a move, John's basement-size SR&RL in On2 shown here has been replaced by a live-steam indoor-outdoor version.

3

It pays to regard engine terminals as "industries." Coal or fuel oil and traction sand need to be delivered, and cinders from coal-burning locomotives need to be hauled away, left. A decoder in a motorized Walthers Burro crane, top, provides another job for yard crews to perform during operating sessions. Also visible are the new diesel fueling cranes to the west (left) of the concrete coal dock.

4

Doug Leffler emulated the practice of ultra-practical short lines by using an old tank car body—complete with lettering—as a diesel fuel supply tank, left. As major railroads gradually dieselized, similar small fuel tanks sufficed. The twin tanks at Frankfort, Ind., right, were sufficient for yard switchers and passenger diesels passing through town. When road switchers arrived, six fueling cranes, unloading cranes, and two large fuel storage tanks were installed. *Doug Leffler; NKP photo courtesy Jay Williams*

5

Allowing breathing room between more intensely detailed scenes—"Spaces between places," as fellow modeler Gerry Leone likes to say—lets the viewer's concentration relax and provides visual separation between unrelated scenic elements. A rural farmhouse is sufficient to add a note of interest to an otherwise unremarkable rural vista on my Nickel Plate layout.

6

These two Michigan barns, photographed in 2016 at Allegan and Battle Creek, would be excellent candidates to model on a transition era railroad, but not in their present physical condition. In the 1940s and '50s they would likely have worn relatively fresh coats of paint. *Two photos: Maynard Mitchell*

In the right background is the C&EI's wing of the brick L-shaped depot. On the east side of the double-track main are a pair of shanties, one puzzlingly set facing north-south. Perhaps coal was stored in there, but it could originally have been the Curtis Street watchman's shanty before crossing flashers were installed. Indeed, the NKP shanty disappeared shortly after this photo was taken around 1950, as I don't recall seeing it when we moved to Cayuga from Iowa in 1951.

Out of view to the right is Thompson's grain elevator. I remember corn cobs flying down a pipe into a pit with a raised berm; the cobs were then burned in the open air with no protective screening. Sure enough—there's that berm and pipe to the right of the shanty. It's weed-grown, which must mean it's not yet time to harvest corn.

I'll surely find or confirm more details about this crossing, which is a prominent feature of my HO model railroad, as I again look at this snapshot in the future. But you get the point: What the photographer thought he was recording and what we now see can be very different if only we take the time to study, and restudy, every available scrap of photographic evidence.

Size can count—against you

When I first met John Rogers, he lived not far from me, and he had an O scale layout. He had even scratchbuilt a steam locomotive. That segued into an HO layout set in New England, which in turn gave way to a basement-size On2 railroad depicting the fabled Sandy River & Rangeley Lakes, **2**, and the Maine Central's interchange with it. He now lives in Montana but is still modeling a Maine two-footer, but this time in large scale employing live steam. A recent photo showed a 2-6-2 pushing real snow off the tracks!

John is therefore somewhat of an authority on modeling in a wide range of scales and eras. "Think about how scale affects the era and prototype you choose to model," he advises. "Modern railroading is about large trains. That's not so easy in O.

"I also think we get conceptually closer to transition era-size railroading," he adds. "Psychologically, we can relate more readily to the less-complex world of that time. This is a bit of an art form, and the canvas of the transition era is easier to paint on."

Feeding the iron horse

Steam's basic needs were fuel—typically coal in the East and Midwest and oil or coal in the West—water, traction sand, and lubrication. Coal arrived from on- or off-line sources in open hoppers; I recall seeing a photo of an NKP hopper loaded with coal at the Rutland's coal dock in its namesake city. That makes sense, as there are no bituminous coal deposits in New England, so including a coal mine on a freelanced New England railroad would not make sense.

Coal was dumped into pits and then hoisted aloft by conveyors or small bins on rails. Coal docks were rated by their capacity in tons, which varied from under 100 tons up to several times that quantity depending on the number of locomotives serviced per day. (Matching coal dock capacity and inbound hopper deliveries to actual needs on a model railroad is discussed in Chapter 9.)

For oil-fired locomotives, the process was similar to that used to fuel diesels. But steam locomotives could burn a much heavier (and cheaper) grade of petroleum than their internal-combustion successors. Steam piping in the tender was often needed to keep the Bunker C oil from congealing in the colder months.

Sand arrived in gondolas and was shoveled or scooped up by cranes and deposited in bins. It was dried and elevated, then piped down into the sand domes of waiting steam locomotives, **3**.

Short lines made do with less. Coal was often shoveled into tenders from an elevated bin or adjacent gondola or dumped in using a front-end loader. Sand was carried up to the sand dome in buckets and poured in. Labor was still cheap in the transition era.

7

Two-story depots were typically found in small towns and remote areas where it would be hard for the agent to travel to work from a house in a nearby town. In other cases, it was a benefit of employment. The agent and his or her family lived upstairs. Here a Charles City Western steeple cab rumbles by the Rock Island depot at Marble Rock, Iowa, in November 1963. *Paul Dolkos*

For coal-fired engines, getting rid of burned coal in the form of cinders was not a trivial undertaking. Hopper or gondola loads of cinders were "shipped" each day, **3**, making cinder removal a good "industry" on a model railroad.

These inbound and outbound loads combine to make an engine terminal one of the most important "industries" we can model.

Accommodating change

Modeling the transition era actually means modeling two eras—steam and diesel—as facilities for one mode of locomotion segued to accommodate the other. As the diesel made inroads, its servicing facilities grew from a small fuel-storage tank or two that were adequate to refuel a few diesel switchers, **4**, or maybe even a local petroleum dealer's truck pumping diesel fuel directly into the locomotive's fuel tank, to huge storage tanks that could replenish the needs of an entire fleet. That required a new track alongside fuel unloading cranes.

The new facilities had to be built alongside the extant steam facilities, or entirely new diesel shops and servicing facilities had to be erected—somewhere. And if there's anything diesel locomotives abhor, it's dirt in their inner workings. The grungy, soot-coated old roundhouse was not their preferred habitat.

Diesel facilities can consume more precious space in our meager steam terminals. Fortunately, if we're modeling a specific prototype or basing our freelanced model railroad's engine terminal on a specific prototype using the Layout Design Element approach—building visually and operationally recognizable models of specific prototype locations—then we can do our initial planning with the needed facilities clearly in mind. No retrofitting, as on the prototype, is required; our time line has already marched past that juncture.

Photos **3** and **4** of the Nickel Plate Road's transition era steam and diesel servicing facilities at Frankfort, Ind., which I modeled in HO scale, illustrate the problem—or, seen in a more positive light, model building and operating opportunities.

8

The last roundhouse built in the U.S. was for the Nickel Plate's new engine terminal in Calumet, Ill., just south of Chicago, in 1951. By that time, many long-standing roundhouses were in poor repair or abandoned, but the NKP had taken delivery of ten new Berkshires two years earlier. This facility replaced one built in the late 1800s closer to downtown. *Courtesy Jay Williams*

Structure texture

Scenery gurus Dave Frary and Bob Hayden like to talk about "texture," which applies to scenery, structures, details, and everything we model. The idea is to avoid blandness that bores the eye. We never have enough space to model much of the world at large, so we need to do more with less, all the while avoiding that overcrowded "model display" look.

Fortunately, the transition era almost leads us by the hand in this regard. Texture in the form of ideal modeling candidates was everywhere and especially right alongside the tracks: section houses, interlocking towers, depots, coal docks, water towers and standpipes, roundhouses, even outhouses. They're not only a scratchbuilder's delight but are also amply represented as kits and ready-built offerings in most popular scales from a wide range of manufacturers.

This is true whether we choose to model the mighty Pennsylvania or Union Pacific or one of the hanging-on-by-a-thread branch, short, or narrow-gauge lines. *Model Railroader* once published a classic cartoon by Hal Kattau showing a historical marker that read: "On this site is the only building in the State of Colorado that

9

Livestock holding pens such as this one at Middlebury, Vt., on Randy Laframboise and Mike Sparks' HO tribute to the Rutland Railroad were still common in small towns during the transition era. That was largely curtailed by the late 1950s and '60s as most livestock was being trucked to slaughterhouses. *Randy Laframboise*

10

Rutland Pacific no. 81 on northbound milk train no. 87 spots an empty milk car (a roll-off tank on a flatcar) at the Sheffield Farms creamery in Ferrisburg, Vt. The Tarvia tank car on the team track at left carries a load of asphalt. The milk business was gradually lost to trucks. *Randy Laframboise*

has not been offered in kit form."

So our main task is to not get greedy. Less is more—our models need room to breathe. Just as it's easy to cram in too much track, it's also far too easy to stuff six buildings into a space where one would be more effective, **5**.

It's also easy to over-weather structures. The transition era is far removed from today—more than a half-century ago. So when we drive along the right-of-way of a main line we plan to model as it appeared in the '50s and come across a barn that has faded paint, missing boards, and a sway-back roof, **6**, we need to remember that this is not how it appeared back then. In such cases, excessive texture is not our friend.

Two-story depots

Most small-town depots were wood or brick one-story affairs. They comprised an office for the agent, a freight-storage area, and a passenger waiting room. Originally, there may have been separate waiting rooms for ladies and for gentlemen; in the South, depots, like some passenger cars, had segregated sections.

But we occasionally come across two-story depots, especially in small towns and remote rural areas, **7**. Because there was no nearby housing available where the station agent and his family could reside, the railroad provided an upstairs living space. This extra expense was offset by the added convenience of having someone at hand around the clock in case of emergency. As long as there was a need for agents, there was a need for depots of this type.

Lineside structures

As a youth, I probably spent more time in my hometown's interlocking tower than I did at the depot across the tracks. That's where train orders and messages were issued and the switches for trains entering or leaving the passing track (Nickel Plate Road) or crossing over from the north- to southbound main, or vice versa (Chicago & Eastern Illinois), were lined. Track circuits announced an approaching train with a loud "Ding!" This was indeed where the action was! And it's a function we can model, as discussed in Chapter 9.

Changing times

The 1950s marked the end of the steam era, but there were instances of modernization even for the Iron Horse. In 1951, just two years after its last order for ten more Berkshires from Lima Locomotive Works, the Nickel Plate Road built the country's last roundhouse and turntable at the modern steam terminal on the south side of Chicago in Calumet, Ill., **8**. Depots and other lineside structures were still getting fresh coats of paint.

The importance of the depot itself was on the decline. Railroads that had high traffic densities installed CTC, which eliminated the need for train-order offices at frequent intervals. Better roads meant that one traveling agent in a pickup or panel truck could cover several agencies.

Livestock continued to be moved by rail into the 1950s, so rural towns often

11

Wayne Sittner models the Northeast, where home delivery of coal, especially anthracite, was common throughout the transition era. Wayne utilizes Sylvan, Roco, and Classic Metal Works chassis, scratchbuilds the dump beds, then adds details such as a shovel, broom, and the chute that guides the coal into coal bins. Those modeling a more modern era may find that old beds were simply attached to new truck chassis; the Blaschak truck is based on a Lindberg kit. *Four photos: Wayne Sittner*

12

Jackman Fuels at Vergennes, Vt., was a typical example of retail coal dealers in the Northeast. It was re-created by Randy Laframboise and Mike Sparks for their HO tribute to the Rutland, which ceased operations in 1960 after a lengthy strike. The high doors suggest that at one time coal was shoveled directly into them from gondolas. *Randy Laframboise*

13

Team tracks served as industrial tracks for local industries that didn't have their own sidings. Here two laborers from A. Emilo's Building Supply take a breather after loading lumber onto a flatbed on the Rutland's team track in Middlebury, Vt.
Randy Laframboise

had holding pens and loading ramps, **9**. But most of this once lucrative business was by the late 1950s moving all the way from farms and ranches to slaughterhouses by trucks over those same improved roads.

Creameries were also a booming business early in the transition era, especially in the Northeast, **10**. Again, as highways improved, trucks assumed an ever-greater portion of this business.

Small-town businesses

During the transition era, home heating with coal began to shift to oil, natural gas (in cities), or propane (rural areas). But coal was still a major factor in many areas, so it's appropriate to have a coal delivery truck, **11**, spotted near a low window or small access door in a home's foundation.

Chuck Geletzke recalls that his family's home in Toledo, which was a Sears Craftsman Home constructed in 1932 by a former NYC conductor, had a coal chute leading to a basement coal bin. "The last retail coal dealers that I recall in the Detroit and Toledo area were still in operation on a marginal scale in the late 1980s."

In small towns, coal typically arrived in gondolas rather than hoppers and was shoveled by hand into coal bins. These distinctive structures had doors high on the track side and were located close to the siding, **12**. Labor was cheap and machines were expensive.

Team tracks, named for the teams of horses that used to bring wagon loads of items to be shipped by rail, or haul them away, remained a fixture in transition-era towns and even cities, **13**.

Although new homes, many of them ranch-style, were being built in the 1950s, the predominant exterior coating was still durable white paint. Ensuring that most residential structures are painted white will go a long ways toward evoking the look of this decade, **14**.

Crops

We don't have the space to delve into the agricultural scene in any depth, but one key point bears mentioned. Through the 1950s, most cultivating—removing weeds in fields of crops—was done mechanically, not by treating with chemicals. Unlike today's fields where plants are planted very closely together, in those days crops such as corn and soybeans were planted in rows that were spaced sufficiently far apart to allow the tractor and machinery it pulled to move down the rows without crushing the plants they were trying to protect, **15**.

How far apart were those rows planted? Why, just use your model John Deere or Farmall tractor as a gauge! There are several good sources of model farm machinery, especially in HO, including Walthers, Athearn, and GHQ, **16**.

Mergers and falling flags

As the curtain dropped on the transition era, signs of major change were already apparent. Powerhouse railroads like the coal-hauling Virginian with its fleet of General Electric's latest streamlined electrics and motors that looked like diesel hood units ("road switchers"—see Chapter 4) were merged into competitors—VGN into the Norfolk & Western in 1959. The locomotives drawing power from overhead wire that had relieved steam crews from near asphyxiation in tunnels were no longer seen as saviors in the diesel era.

14

Plain Jane in her durable white coat was still the leading lady when it came to exterior paint on houses. Pastels were making inroads, but the durability of "white lead" paint was a lesson not lost on thrifty home builders and painters through the 1950s. A coat of white paint allows an Atlas factory-assembled house, or several of them, to fit unobtrusively in any setting.

15

Rows of soybeans abut the Nickel Plate's St. Louis Division in southwestern Illinois, top, as three GP9s hustle New Haven and Missouri Pacific piggyback trailers—one trailer per flatcar in those days—westbound. Note the distinct spaces between rows of beans to allow a tractor and cultivator to work in the field without damaging plants. I model bean fields using large brown pipe cleaners covered with green or yellow leaf flakes, a technique developed by Jason Klocke. *Nickel Plate: J. Parker Lamb*

I assembled GHQ John Deere tractor-with-corn-picker and wagon soft-metal kits, then put them to work harvesting stalks of JTT's mature corn. The ears of corn are grass seed.

The catenary supports survived long after the wires came down in the early 1960s, but the spider web of copper cables was gone forever.

The Erie and competitor Delaware, Lackawanna & Western merged just as the transition era ended. But signs of trouble preceded the merger; there was little sense spending money on upkeep when the future was so uncertain.

By the mid '50s, the New York, Ontario & Western was taking its last breaths, its lifeblood of anthracite coal having slowly oozed out of fashion as home heating oil and natural gas sneaked in the backdoor, much as the diesel had begun doing to steam power a decade before. The Rutland was plagued with labor-management dysfunction and barely staggered into the 1960s before calling it quits.

For the most part, however, the transition era was the last hurrah of railroading as fans and modelers came to prefer it. If you liked what you saw in the 1930s, with the major exception of the shiny new diesels and lightweight passenger cars, you'd find the same cast of players in the 1950s: Steam still played a major role; cabooses marked the end of trains; switchmen still "decorated" the tops of the plain Jane boxcars that still delivered most of the non-mineral goods (and some of them as well); mail and out-of-town newspapers arrived by train; the Railway Express Agency (REA), not UPS or FedEx, delivered parcels (see *Express, Mail & Merchandise Service* by Jeff Wilson; Kalmbach, 2016); and the friendly local station agent was still at work to sell you a ticket on a passenger train that, through connections, could take you to almost any city or town on the continent.

No wonder it's such a popular period to model!

CHAPTER NINE

Operational considerations

Train movements governed by timetable and train-order rules, which on model railroads ideally employ a dispatcher and one or more operators, were common through the 1950s and into the 1960s and beyond. Here Jeff Ward sits at the Frankfort operator's desk on my model railroad; he also handles orders and messages for three other towns. A second operator handles the four towns in the other main aisle. The operators "OS" trains to the dispatcher as they pass each station; road crews normally talk only to operators. *Mike DelVecchio*

We've discussed the way things were during the transition era, and we've taken a closer look at locomotives. Now we'll explore ways to use those exquisite models more realistically. We need not strive for individual showpiece models but rather for using reasonably representative miniatures to re-create railroading as it was during this time period. Plausibility is the watchword. Everything from choice of prototype to detailing to paperwork to operating scheme, **1**, should rise to, but not necessarily go beyond, the level of Good Enough.

What went by rail

Almost everything! As today, trucks hauled a wide variety of items between cities, but the lading that traveled in boxcars was astounding. The reasons were many, especially in the pre-Interstate days when every small town had a varied list of small "industries."

Minneapolis & St. Louis modeler Clark Propst compiled a list of just some of the products shipped by rail to tiny Dallas Center, Iowa (1950s population: 1,000) in the early 1950s. The summary of this list in **2** contains some interesting information. Not including shipments to the M&StL's local agent and other railroad employees, I count nine different consignees. Compare that to most small towns today: no local industries served by rail, and probably no railroad at all. That alone accounts for much of the popularity of this era for modelers.

Most of the products sent to Dallas Center aren't surprising: lumber, feed, farm implements, coal, cement, fertilizer, molasses (to be mixed with feed), furniture, twine, distillate, and company supplies.

"Distillate" is a product that we don't think much about today. Mont Switzer, owner of Switzer Tank Lines in central Indiana, notes that "With the exception of Jet A, K-1 is the most refined of all distillate classes. Only small quantities of K-1 were used for lighting—hand, barn, and barricade lanterns. Like many rural areas, Henry County, Ind., was completely electrified by 1940.

"No. 1 furnace oil and K-1 were pretty close to the same thing and offered the cleanest burn for home heating," he adds. "That's probably where most of the action was. In the Midwest, many folks converted from coal to oil heat right after WWII."

Back to Dallas Center, Iowa: Note where the products in Clark's list came from. Especially colorful origins are oyster shells and molasses from Louisiana and limestone from Weeping Water, Neb. The coal logically came from southern Illinois, not Appalachia. Salt, perhaps for cattle licks, came from Kansas in a Chesapeake & Ohio boxcar and Louisiana in an

Dallas Center, Iowa, M&StL station agent records in 1953-55 • DATA COLLECTED BY CLARK PROPST

RR	TYPE	NUMBER	CONTENTS	ORIGIN	RECEIVER
ACL	Box	22750	Lumber	Lacoochilt, Calif.	Hoover Lumber
ATSF	Box	139611	Cement	Des Moines, Iowa	Hoover Lumber
IC	Hpr.	68184	Coal	West Frankfort, Ill.	Farmers Co-op
ATSF	Box	148539	Feed	Omaha, Neb.	Reason Produce
B&M	Box	75498	Lumber	Madras, Ore.	Hoover Lumber
B&O	Gon	252939	Pipe	S. Lorain, Ohio	No. Natural Gas
B&O	Box	274684	Mdse.	Cloquet, Minn.	Dallas Ctr. agent
B&O	Box	280762	Lumber	Potlatch, Idaho	Hoover Lumber
C&O	Box	4649	Feed	Omaha, Neb.	Reason Produce
RI	Box	148773	Bran	Des Moines, Iowa	Farmers Co-op
C&O	Box	27337	Stoves	Gaird, Ala.	Weather station
C&O	Box	80662	Implements	Ottumwa, Iowa	Wall Implement
NH	Box	35081	Fertilizer	S. Omaha, Neb.	Farmers Co-op
C&O	Box	291523	Salt	Hutchinson, Kan.	Dallas Ctr. agent
NYC	Hpr.	872752	Coal	Harrisburg, Ill.	Farmers Co-op
CGW	Box	91089	Oatmeal	Des Moines, Iowa	Farmers Co-op
C&NW	Box	80186	Oil meal	Minneapolis, Minn.	Farmers Co-op
NP	Box	19293	Bran	S. Omaha, Neb.	Farmers Co-op
UP	Box	198312	Alfalfa meal	Josslyn, Neb.	Farmers Co-op
GN	Box	47479	Midds	Grand Forks, N.D.	Farmers Co-op
I-GN	Box	9776	Salt	Jeff. Island, La.	Farmers Co-op
MP	Hpr.	59821	Coal	Crescent, Ill.	Farmers Co-op
NYC	Box	157732	Fertilizer	Houston, Texas	Farmers Co-op
MKT	Box	95620	Co. material	Albert Lea, Minn.	Farmers Co-op
CGW	Box	89082	Lumber	Gardiner Jct., Ore.	Hoover Lumber
L&A	Tank	4247	Molasses	Gramerey, La.	Farmers Co-op
CN	Box	476135	Lumber	Attiabasco, Alberta	Hoover Lumber
CN	Flat	661995	Generators	Midwest City, Ore.	Resident engr.
CNJ	Gon	1897	Pipe	Galewood, Ill.	Iowa Pwr. & Lt.
CNJ	Box	20771	Wire	Steelton, Minn.	Hoover Lumber
C&NW	Box	8354	Oyster shells	Morgan City, La.	Reason Produce
CP	Box	251700	Lumber	Port Hammond, B.C.	Hoover Lumber
D&RGW	Box	67843	Furniture	Grand Rapids, Mich.	Metropolitan
MP	Box	30016	Limestone	Weeping Water, Neb.	Farmers Co-op
KCS	Box	26042	Posts	Joplin, Mo.	Hoover Lumber
L&N	Box	39909	Prefab bldg	Memphis, Tenn.	Weather station
M&StL	Flat	23053	Rail	Albert Lea, Minn.	Section foreman
M&StL	Gon	30005	R.R. ties	St. Louis Park, Minn.	Section foreman
M&StL	Box	53058	Grain doors	New Ulm, Minn.	Dallas Ctr. agent
M&StL	Cvd. Hpr.	70171	Cement	Des Moines, Iowa	New Ulm, Minn.
MILW	Box	27355	Twine	Brantford, Ore.	Farmers Co-op
SOU	Gon	60502	Co. scrap	Twin Lakes, Iowa	Section foreman
UTLX	Tank	18295	Distillate	Des Moines, Iowa	Blandard Oil
WM	Box	28243	Steel	Canton, Ohio	Hoover Lumber

2

Wood grain doors were made from multiple boards nailed into a panel. The panels were stacked and nailed across a boxcar's door openings to seal the car for loading grain. Spare grain doors were typically found at grain elevators, as on Bill Darnaby's Maumee Route. Note the owning railroad's initials on each door; they want them back! *Bill Darnaby*

International Great Northern boxcar, which was sort of on their way back to home rails per car-service rules.

One item that caught my eye was grain doors. These were panels, typically comprising three wide boards nailed together, that were in turn nailed to the inside of boxcars being loaded with grain. The name of the railroad was stenciled on each panel, and that railroad wanted them returned after the car was unloaded at a distant elevator or other facility. That one car may have contained enough grain doors to supply several elevators in more than one town, thus providing an interesting operational opportunity for a Granger Belt model railroad. They also make good scenery props near an elevator, **3**.

Before UPS and FedEx

In the days before brown or white step vans regularly pulled up to our doors to drop off parcels, we typically received everything that was a bit large for the post office to handle—appliances, furniture, car batteries and other auto parts, lawn mowers, bicycles, and so on—by rail.

It could arrive in a boxcar coupled behind a local freight's locomotive, **4**, or be in an express or baggage car of the Railway Express Agency. For more information on this fascinating topic, see Jeff Wilson's book *Express, Mail, and Merchandise Service* (Kalmbach, 2016).

This impacts transition era operating sessions. In addition to doing any in-town switching, the local should stop in front of the depot's freight door to allow the agent to roll a baggage wagon up to the boxcar door so the LCL (Less than Carload Lot) freight could be unloaded. For passenger trains, there may be REA packages to handle from the baggage car.

Sounding off

The advent of sound decoders offered many operating opportunities. Locomotives' whistles warn of their approach and convey information from the head end to railroad employees elsewhere. Bells have a similar warning function.

So we began to learn whistle signals—two blasts to acknowledge a highball from the conductor before releasing the air brakes and opening the throttle; two shorts to acknowledge other signals; three shorts before backing up; a long and two shorts to warn trains we met or passed that we were carrying green flags or classification lights (not markers; they're on the caboose or last passenger car) and hence at least one more section would be following in our wake, **5**. We could even use the whistle or horn to send out or call in a flagman when we stopped on the main, a subtle reminder to a fellow crewman (or ourselves) that, even if we're a scheduled train, we do have to protect the rear against the approach of

4

On St. Louis Division locals, the Nickel Plate coupled a rider car behind the engine to handle Railway Express Agency shipments. The "rider" was paid equally by the NKP and REA. Behind it was a boxcar loaded with less-than-carload-lot (LCL) parcels. A train crewman and the station agent unloaded that car as it stopped by every depot along its route.

another train by sending out a flagman per Rule 99, **6**.

Note that unscheduled or "extra" trains carry white flags or class lights to announce their status, **7**.

Dynamic brakes worked by turning the locomotive traction motors into generators driven by train movement, and the current they developed not only created resistance against the axles turning but also generated a lot of heat. That's why you can see slotted openings on the roof behind the cab of dynamic-brake-equipped Electro-Motive FTs, F2s, and F3s and exhaust fans on later F units. GPs and SDs with dynamics had exhaust fans and air-intake "blisters" centered atop the long hood. Alcos had dynamics brakes tucked into the short hood of RS-2s and -3s and a series of boxy shapes atop the long hoods of RS-11s. Many diesel sound decoders respond to a function button that turns the howl of the dynamic brake exhaust fans on and off.

5

Green flags mean that at least one more section of the train is following behind. Failure to notice the flags (or green classification lights) could result in a head-on collision, so the crew "carrying the green" had to call attention to the flags with a long and two short whistle blasts—and receive an acknowledgement from the opposing train.

The local's flagman has lit a fusee to warn trains approaching from behind that the local is fouling the main track ahead. This tiny red flare (Fusee Pro), made by Logic Rail Technologies (logicrailtech.com), actually flickers and is actuated by pressing a button on the fascia. Burn time is variable; I set it at 10 actual minutes.

White flags or classification lights (not "markers"—those are at the rear of the train) as displayed by Monon RS-2m no. 59 denote an extra (unscheduled) train. Scheduled trains need not be aware of extras. It's up to the extra's crew to stay out of the way of scheduled trains unless the dispatcher issues orders to the contrary.

Coaling model steam locos

I don't recall ever seeing anyone figure out how to gradually reduce the pile of coal in a steam locomotive during an operating session, or to refill it later on. But Bill Darnaby showed me a way to simulate this everyday operation.

Following his example, I made up small slips of paper marked "10 TONS COAL", **8**. I put five of them in a twin hopper destined for one of my two coal docks, seven in a triple hopper. During an operating session, the roundhouse foreman is charged with removing the coal load from each hopper as that hopper is moved by hand into the coal-dumping shed. The associated five or seven slips are then put into a bill box marked "COAL."

As he coals each locomotive by pressing a button marked "COAL," which turns on a sound module that simulates dumping coal into a tender, he moves one (for small tenders) or two (for large tender) slips from the COAL bill box to another box marked "USED COAL."

As his stash of coal in the coal dock is depleted, he has to unload another hopper. If he runs short of loaded hoppers, he informs the yardmaster that more coal has to be switched into the coal dock delivery track.

Poling and dropping cars

Railroading became increasingly safer for employees as the decades wore on, but there were still some "interesting" practices that survived into the late-steam era and beyond. One of them was "poling" cars. A long wooden pole secured below the deck on a modern steam locomotive's tender, **9**, indicates management hadn't banned the practice.

One end of the sturdy, metal-tipped pole was inserted into a pocket on the pilot of a steam or diesel locomotive and the other in a similar pocket in a freight car on an adjacent track. This allowed the crew to move that car out of the way in case it, for example, fouled a switch after a "drop."

Crews dropped cars into a facing-point siding, thus avoiding a runaround move, by accelerating the locomotive and car or cars to be dropped to walking speed, cutting off the cars, and then racing ahead to clear the siding switch. Once the locomotive had cleared the switch points, a switchman threw the switch for the siding, and with any luck at all the cars rolled off the main and into the siding. A brakeman was riding the cars and wound up the handbrake to bring the cars to a stop at the desired spot.

Not hairy enough for you? Try a Dutch drop. Here the engine got the car rolling, and it was again cut off. But the car's destination was behind the engine, so the engine had to clear a switch, stop, wait for the switch to be thrown, reverse its direction and race toward the oncoming cars, and—again, with any luck at all—clear the entire turnout so the switch could be thrown to allow the car to roll past and onto the track the locomotive had just vacated.

The late John Armstrong equipped some O scale boxcars with chain-driven flywheels that allowed cars to be kicked into yard tracks and could presumably be dropped as well. The flywheels also stored enough momentum when rolling along in a train to aid and abet some spectacular derailments, John told me.

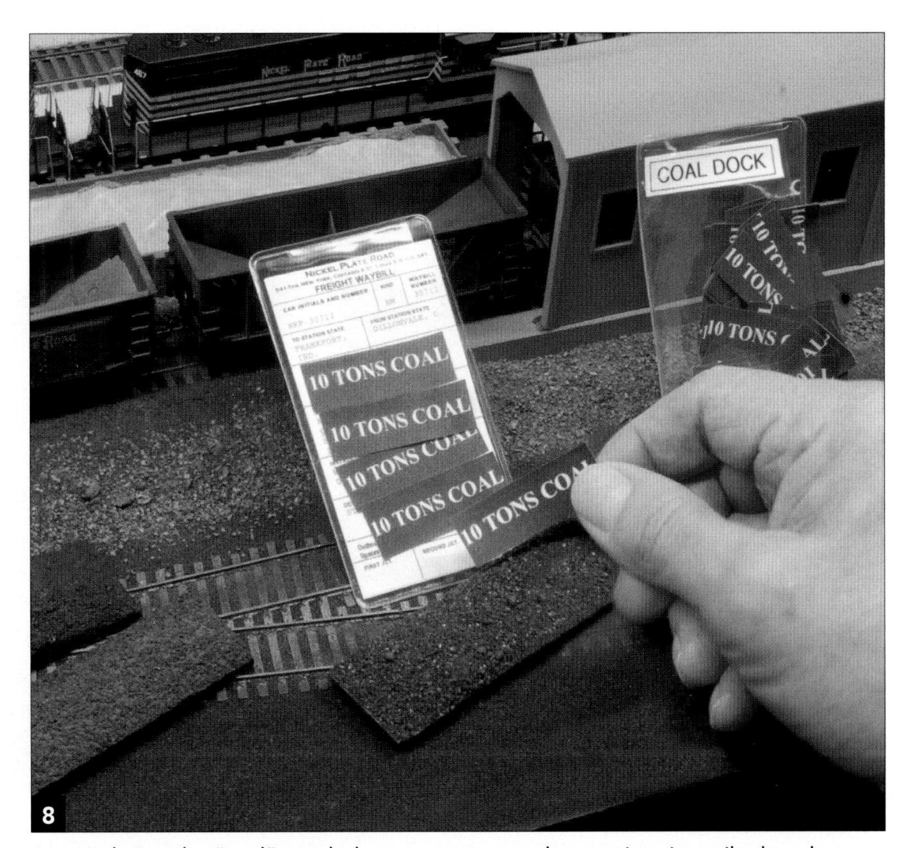

Accumulating the "coal" needed to operate steam locomotives is easily done by creating 10-ton coal slips. Five or seven are inserted into the waybill sleeve for 50- or 70-ton hoppers. The slips are "unloaded" at the coal dock as loads are removed to empty each hopper. One slip is used to coal a small engine.

Shorter cars

On our space-constricted model railroads, every inch counts. That provides another advantage of modeling the transition era: shorter cars. The 50-foot double-door automobile-carrying boxcar dates back to the 1930s, but the 40-foot boxcar was still the mainstay of the fleet well into the 1950s. Four 50-footers are equal to five 40-footers, which makes trains look longer.

Similarly, the 70-ton triple (three-bay) or quad (four-bay) hopper car was no stranger during the war, but the 50- and 55-ton twin hopper was still the most common type of coal carrier. A train of twins looks longer than a train of identical-length longer cars.

Wood 36-foot refrigerator cars (reefers) remained popular well through the 1950s for two reasons: Wood is a good insulator and won't rust; and many loading and unloading sites were based on their 36-foot length. The cost of replacing them wholesale wasn't justified. What did them in was mechanical cooling; icing long strings of reefers wasn't anyone's idea of efficient operation. Mechanical cars, however, did not make major inroads until after the transition era.

The team track

Practically every town had a team track, so named because teams of horses pulled wagons alongside the track so boxcars could be directly unloaded. This was especially handy for small industries that shipped or received lading by rail but didn't have their own siding.

Even industries such as lumberyards that had their own siding might use a team track to avoid switching charges. Bert Kram, who worked at Kramer Bros. Lumber in Frankfort, Ind., in the late 1950s, recalls that "Kramer's was then served by a siding from the Pennsylvania Railroad and received one or two boxcar loads per week. But they also received loads via the nearby Monon team track. Kramer had two smaller retail sites in adjacent towns. Those sites did not have rail sidings; instead, portions of the materials received at Frankfort were trucked to them.

"Among the products received were dimensional lumber in lengths of 6 to 16 feet, 4-foot sheets of plywood of various thicknesses and lengths of 8 to 14 feet, plasterboard in the same width and lengths, asbestos wallboard, lath in bundles for plaster walls, packages of roofing shingles, bags of plaster or lime (not mixed in one car), and cement in bags that I recall weighed 94 pounds each!

"All of this came in boxcars; we never received flats," Bert recalls. "Lumber was manhandled out of the car and loaded onto 'yard buggies,' which were then moved by a forklift."

Loading livestock

Gordon Locke recalls that "Back in the early '60s I would catch a caboose hop out of Temple, Texas on the Santa Fe

That odd-looking pole carried under the tender of some steam locomotives, above, was used to pole, or push, cars on an adjacent track. Yes, it was dangerous, but freight cars and even some second-generation diesels, right, still sported poling pockets for just that purpose.

9

about 5:00 a.m. We'd run to Lometa, pick up 60 empty stock cars and continue to San Saba on the San Saba District branch to the Owens Brothers Ranch and load a train of yearlings for Colorado summer pastures.

"This was done with a portable loading chute, so we did not have to spot each car. I believe it took 5 or 6 hours to load the train, which works out to 5 to 6 minutes per car."

Monochrome boxcars

For those put off by the general lack of color in the freight car fleet, consider how this adds to operational flexibility. Highly distinctive cars such as the Bangor & Arroostook's red, white, and blue State of Maine boxcars stand out like a neon sign, **10**, whereas a boxcar or Tuscan red car sort of blends in.

So if the same car shows up at a local industry more often than it probably should, be sure it's a plain Jane boxcar, not that rolling billboard.

Loading homebound reefers

Improving car utilization is simply a matter of not moving cars filled with air. So shippers and railroaders alike look for creative ways to load an empty car while still sending it back to or toward home railroads. This included ice reefers.

Meat reefers (most of which were privately owned) were not reloaded with general cargo, but those that had come north or east with fruit or vegetables in cartons would be ideal to reload with canned goods, as their insulation would offer some protection against freezing.

RBLs, the Association of American Railroads designation for insulated plug-door boxcars without ice bunkers, were beginning to show up in the mid-1950s. Barry Karlberg recalls a more recent experience in Wisconsin where RBLs were warmed up with hoods that enclosed the plug door. The canned goods were loaded and secured, and the plug door locked and sealed. He was told that the water in the cans would stay warm enough to keep the contents from freezing on loads headed south.

Although the term "refrigerator car" suggests keeping its lading cold, portable heaters were also installed in reefers to maintain warmth during the winter months. Karl Schoettlin recalls the Santa Fe offering heaters under "protective service" into the 1980s.

The advent of piggyback

By late 1954, railroads were already experimenting with trailers on flat cars, or TOFC, **8-12**. So if trains including lots of piggyback flats are a major criterion, consider setting your railroad in 1955 or later.

The original pig flats were nothing but flatcars with hitches and chains to secure the trailers. Guide rails along either side kept the trailers aligned with the car centerline as they were driven on board "circus style." Cars were commonly 50- to 53-foot converted flatcars.

By the late 1950s, specially built flatcars that would hold two longer trailers were coming on line. But the New York Central's innovative Flexi-Vans and the 89-foot TOFC flats as well as bi- and tri-level auto racks and high-cube boxcars would have to wait until the next decade.

Car inspectors

Every car that entered a freight yard had to be checked over by a car inspector (or "car knocker," a colorful term derived from the inspector's banging on parts to be sure they were securely attached and flipping journal

10

Which car in the yard caught your eye? Which one will you remember seeing session after session? That suggests we should avoid routing brightly painted cars to local industries on our railroads, as crews will too easily remember seeing them during their last cycle—unless you're modeling a railroad that had such cars on its roster.

box lids open to check the lubricant and cotton waste that wicked oil onto the axle journal's surface).

This is another of those jobs too few modelers consider, but it leads directly to modeling another "industry"—that is, a place to spot and pick up freight cars. In this case, if your yard crew sees a car with a missing brake wheel or door track, loose running board (not "roof walk"), broken ladder, or missing grab iron or step, the car can be ticketed as "bad order" and routed to your RIP (repair in place) track. The car stays there until the owner fixes it and returns it to the RIP track. There it can be picked up and continue to its original destination.

Assigned engines and cabooses

Back in steam's early days, a locomotive was often assigned to a specific engineer. By the transition era, however, this was impractical. That said, on many short lines and branch lines the same engine was often assigned to the same crew out of sheer practicality.

Cabooses were a different story, as they were often the crew's home away from home even well after the 1950s. As an example, Nickel Plate caboose 1149 was usually spotted by the yard office in Michigan City, Ind. I became good friends with conductor Bill Love. His home base was at the other end of this division in Peru, Ind., so he stayed overnight in the wood caboose. He had the engine crew spot the caboose next

11

When cabooses were the conductor's home away from his home terminal, they were typically assigned. Later, cabooses were still assigned to divisions. On my railroad, cabooses that come in to the sprawling yard at Frankfort, Ind., from one division have to be cross-yarded from the eastbound to the westbound yard caboose track, or vice versa. The arrows show the respective caboose tracks.

to a telephone pole where he had rigged up a TV antenna and AC power that he could easily plug into.

I don't assign cabooses to individual conductors, but I do assign them to divisions. That requires a caboose that enters Frankfort's eastbound yard from the west to be cross-yarded to the westbound yard's caboose track, **11**, and vice versa. It can then return west to the next division point in Charleston, Ill., and the cycle then begins anew. But even in yards with only a single caboose storage track, assigning cabooses to specific divisions adds a realistic operating complication.

The "hog law"

During the transition era, railroaders could work a full 16 hours per day. If you use a fast clock and run operating sessions longer than 16 hours, it would be possible for a crew to "outlaw" or go "dead on the law." They would need to anticipate this and get their train into the clear before their duty time had expired and notify a dispatcher to have a crew "dog-catch" their train.

My operating sessions are 12 fast-clock hours long, so no crew could possibly run out of time. But the dispatcher could consider the law in terms of each train, not each crew. He has the on-duty time on his train sheet. So when the next session begins and several trains are typically "frozen in time" in their runs somewhere on the division, he could notify the crew that dog-catches them how many hours of service they still have to work. That would cause job assignments to cycle and "employ" more road crews.

Telegraph communications

By the 1950s, communication was often by telephone, but the telegraph was still in daily use. One of my prized possessions is the telegraph key used by the second-trick operator in Cayuga, Ind., a gift from Charlotte Schwab Miller, daughter of first-trick operator Bill Schwab.

When I broke in as an operator in the mid 1960s, my mentor, Bill Yuill, still used his "bug" to chat with other veteran "ops" on the NKP's Peoria Division. But I never developed the skill to send or read the clicks that comprise railroad Morse, which is considerably more difficult than the beep-beeps of International Morse Code that many of us learned as Boy Scouts.

On Andrew Dodge's former On3 Denver, South Park & Pacific, train crews used International Morse code to tell the dispatcher they have arrived or left a station. On his new Proto:48 (O fine-scale) Colorado Midland, he still uses the "beep" keys but now uses conventional Railroad Morse.

"I looked at the railroad code used, and it was a lot simpler," Andrew reports, "especially with the numbers. I am bending the true nature of it a bit, but the dots and dashes are being sent in a similar manner as International Morse code. I cannot replicate the clicks and spaces of a telegraph key, but this is another step closer to historical accuracy."

Andrew prepared a "cheat sheet" for the operators and dispatcher, **12**. "I have had one operating session so far on the CM and had to make an adjustment: a return to calling the dispatcher when arriving and departing. I thought I could drop the departing message using timetable and train-order operation, but TT&TO authority Steve King wanted both."

Call Signals

Station Names

Station abbreviations are:

Ark. Junction *(• — | — • — •)*	AJ
Busk *(— • • • | — • —)*	BK
Ivanhoe *(— • | • —)*	NA
Stellar *(• • • | • (short hold) • •)*	SR
Thomasville *(• • (short hold) • | • • —)*	CU
Basalt *(• • • | • — • •)*	SX

Mainline Eastbound Trains

(Eastbound Priority over Westbound Trains)

#6	(— — | • • • • • •)
#8	(— — | — • • • •)
#38	(• • • — • | — • • • •)
#42	(• • • • — | • • — • •)

Sequence of Transmissions

(Provide full sequence of calls each time for arrival or departure)

Crew will Send	**Dispatcher will Send**
Station name:	Acknowledged (AK):
(Watch for light)	AK (• — | — • —)
Train number: (See above for #s)	AK (• — | — • —)
Arrived (O S): OS (• • | • • •)	AK (• — | — • —)
Departing (D T): (DT) (— • • | —)	AK (• — | — • —)

12

Andrew Dodge made a telegraph code sheet that tells train crews how to send an arrival or departure message from a given station. Only the dispatcher needs to be able to "read" the code. *Andrew Dodge*

Moving toward longer trains

Some railroads, including the Monon, took little time to understand that they could run longer (hence fewer) trains by dieselizing their main lines. If you've elected to model one of those railroads, you have a tough decision to make. If long consists of F units are high on your must-have list, you may find that steam vanished almost overnight and co-existed with the covered wagons for mere weeks or even days. What you gain in hardware preferences could be a major loss in operating potential as the railroad sits idle for longer periods between long trains, and meets are few and far between.

I encountered a similar choice in modeling the NKP's St. Louis Division. In one week in July 1955, engineers' time books, in which they recorded which trains they ran, the locomotives assigned to their trains, and other information that's helpful to modelers, showed that the 600- and 700-series steam locomotives faded away as 400- and 500-series Geeps and RS-3s took their place. My initial assumption that I could find a magic time when both steam and diesels peacefully co-existed was not to be. The week the GP9s showed up, steam departed.

I could therefore run GP7s and RS-3s out of the terminal at Frankfort, Ind., but only northwest to Peoria and northeast to Toledo—that is, on short trips into staging—but not to St. Louis or toward Cleveland. So far, my memories of and fondness for Berkshires and Mikados more than offset that loss.

Under timetable and train-order rules, operators would "hoop up" clearance forms, orders, and/or messages to passing train crews on the fly. Replicating the essential elements of this archaic system is very practical on a model railroad and ensures that train crews play an active role in conducting safe and timely train movements. *Wayne Sittner*

Where two railroads crossed at grade—here the Maumee Route and the New York Central at Edison, Ohio—there was almost always an interchange. A single leg of a wye can host a wide variety and quantity of freight cars as a "universal industry." An armstrong interlocking frame controls the junction. *Bill Darnaby*

Most of the interlockings on the Maumee Route employ Hump Yard Purveyance levers connected to electrical switches. The switches provide position information to software on a laptop. Bill Darnaby described how he built these plants in *Model Railroad Planning* 2010. *Bill Darnaby*

Bill Darnaby worked with CTC Components (ctccomponents.com) to build this tower-size Union Switch & Signal CTC machine to control the Maumee-C&O interlocking at Gastonia, Ohio. It's easy enough to operate that either a towerman or a passing train crew can set the desired route. *Bill Darnaby*

Finding out how things worked

I mentioned employee time books. These were pocket-sized record books in which each operating employee recorded his time on and off duty and mileage, and from this his pay could be calculated.

Unlike so many other bits and pieces of railroad paperwork, time books were retained indefinitely by each employee. If you know someone who worked on the railroad you're modeling during the transition era or, more likely, the son or daughter of the professional railroader, he or she may have dad's time books stashed away in the attic. It's worth asking.

Railroad-specific historical societies are likely to be a treasure-trove of information. Many of them have model-oriented print and/or online publications. Back issues are often available to members and nonmembers alike. You can show your appreciation for those efforts by becoming a member.

Dispatching your railroad

The transition era preceded track warrants and other radio-based movement-authority communications that are common today. Centralized Traffic Control (CTC) was well established, but timetable and train-order dispatching was still very common.

What type of dispatching system you employ will depend to a large extent on what your chosen prototype used. Indeed, a strong case can be made for choosing a part of a specific prototype that used the dispatching system that most appeals to you.

If you're freelancing, you have more options. This is not the forum to present a lengthy discussion of the pros and cons of the various types of dispatching methods; for that I refer you to my book *Realistic Model Railroad Operation, Second Edition* (Kalmbach, 2013).

I will point out, however, that understanding the various dispatching options, skill sets, costs, and so on is one of the most critical choices facing the layout designer and builder. Running trains through stunning scenery is delightful at first but can get old in a surprisingly short time. There are those who derive much of their modeling pleasure from building yet another layout. But I suspect that in many cases a design that supported more realistic operation would have added years to the longevity of one of their earlier efforts.

Modeling the transition era cannot be done simply by building models and scenery that faithfully represents that well-documented time. Things happened back then, and we know a great deal about them. So our job is as much about modeling what happened as about the things that events happened to and with.

17

Timetable and train-order operation ideally employs a dispatcher and at least one operator. Here NKP Third Sub dispatcher Ron Von Werder dictates a train order to an operator on my railroad. Posted in front of him is a copy of an actual Third Sub train sheet, which dispatchers can consult as needed.

18

One tower operator controlled a number of junctions on Doug Tagsold's Terminal of Toledo HO railroad. He mounted small video cameras looking both ways at each junction and with two small screens for each junction. *Doug Tagsold*

Paperwork

If you're modeling a small, one-train-a-day short or branch line, you can get by without any paperwork. There's no chance of a collision, so no schedule is required. And you can memorize the spots where each carload of freight is to be delivered or picked up.

As the size and complexity of a model railroad grows, so too do the systems needed to manage its operation. I can't hope to remember what several hundred freight cars and a dozen trains are supposed to do today on my model railroad. Moreover, one person didn't make all of those decisions on the prototype in the 1950s, so I shouldn't be making them, either.

The beauty of modeling a prototype railroad in a well-defined time and place is that history makes those decisions for us. Our job is to uncover enough historical background information to make rational decisions about how best to interpret that in our simulations.

19

The Chesapeake & Ohio dispatcher sat in the second-story bay window of the Hinton, W.Va., depot, above. On Ted Pamperin's World War II-era HO tribute, the dispatcher's office is in a room adjacent to the railroad, right. Ted rigged a small video camera to a large screen to give the dispatcher a similar view of the modeled yard. *Ted Pamperin*

Note that word: simulations. We're not playing games or re-inventing the wheel. Ideally, we're simulating what actually happened back then. And back then they used paper—a lot of paper—to make the railroad operate efficiently and safely.

Since we don't employ legions of clerks to keep track of everything, we need to find systems that do most of the work for us, or we simply ignore aspects of railroading that have no value on a model railroad (billing, for example). But we don't want to toss many job descriptions out with the trash too hastily, as there are some seldom-modeled "jobs" that turn out to be interesting and even fun when adapted for model railroad use.

In a CTC environment, trains moved by signal indication, so schedules weren't required. The dispatcher's ability to "see" the location of every train and decide in real time what to do with it made keeping densely trafficked single-track railroads much safer. In model railroad terms, the corresponding sea of signals is a joy to behold. It places few knowledge requirements on train crews, as they simply obey signal indications and let the dispatcher route their trains.

This is one area where modern computer monitors may have a place on a mid-20th-century model railroad. Building an actual CTC machine can be expensive, but there are software packages that will create reasonable facsimiles on the screen. Nonetheless, the costs associated with signal acquisition and installation requirements can be significant.

In a timetable and train-order environment, **13**, the schedule was the basis for everything that happened. The cost is low: Create a timetable using spreadsheet software and have the local office-supply store print a few dozen.

But the learning curve for crews is much higher than with CTC. If you have a regular crew, as I do, you can expect them to become experienced with TT&TO rules and come to thoroughly enjoy the challenges. It is more difficult to accommodate boomers (visitors), but this can be surmounted by having two-person road crews.

You'll also need a book of rules. If you're modeling a specific prototype, rulebooks of the period can usually be picked up for reasonable prices online or at train meets. And while you're poking around at the swap meet, try to find an *Official Guide to the Railways* for the year you're modeling. These contain a wealth of information about which railroads served which towns, making it much easier to set up plausible routes when you create waybills to route freight cars.

Waybills

Just as trains moved by schedule authority, train orders, or CTC signal indications, freight cars needed a

formal system to get from where they were loaded to where that lading was needed. I devoted an entire chapter to a more realistic waybill format in *Realistic Model Railroad Operation,* so we won't spend much time on that topic here.

The point to take home here is that moving cars according to a system is as interesting as dispatching trains. If we spend time and money ensuring that our models accurately depict their prototypes, moving trains and the cars in them with equal regard to prototype practice seems to be a good investment.

Interchanges and interlockings

In Chapter 1, we briefly discussed how important the interchange between two railroads was during the transition era as a "universal industry" that can accept or deliver virtually any type or quantity of freight car, **14**. As I described in *Space-saving Industries for Your Layout,* there are four different ways to effect interchange: a short track truncated by the backdrop or aisle on which a few cars are interchanged each day; a manually operated interchange that a crew member is assigned to work during an operating session; an automated interchange that uses a detection circuit and relay to deliver another cut of cars after each preceding cut is picked up by your railroad; and a virtual interchange to a disconnected length of track.

I employ a general freight agent to check industries to see whether they will need more empty cars "today." The agent then works with the general yardmaster to ensure that the needed empties are delivered as expeditiously as is practical. He or she can also manage the two manually operated interchanges.

Interlocking towers were ubiquitous through the 1950s. Some busy or remote crossings were automated by then, but many were still guarded by a towerman working a system of levers to set up and clear a route through the "plant."

Using levers such as those produced by Hump Yard Purveyance (humpyard.com), **15**, it's practical to create one or several working interlocking plants and to assign their operation to a towerman or perhaps to an agent-operator who also copies and posts train orders and messages.

Some towers and depots were equipped with mini-CTC machines, **16**, that were operated by the leverman or station agent, who also served as a telegrapher. This is a very practical way to enjoy working with a CTC machine while the rest of the railroad is dispatched using timetable and train-order rules.

Just remember that most such jobs require a workspace such as a desk. We need to allow room for the desk and chair, not to mention a dispatcher's "office," when designing our layouts, **1** and **17**. Dan Holbrook provided a compact area decorated to look like a corner of a caboose with a desk for his road crews to convert waybills to switch lists.

Since we can't always locate our operator(s) —I have two—where they can easily see all of the stations they're assigned to manage (four each on my

20

This Rutland work train includes a Ledgerwood ballast spreader. A plow was dragged through the ballast cars by cable, pushing ballast out the side doors. Rutland modeler Randy Laframboise notes that the prototype had steel plates between cars so the plow could run continuously thru all the cars. The car next to the engine has a winch that was used to pull the plow. He plans to dispatch this work train during operating sessions. The ballast cars are resin kits from Funaro & Camerlengo; Randy scratchbuilt the Ledgerwood car. *Randy Laframboise*

NKP), tiny video cameras mounted in lineside structures or hidden in trees or scenery can relay sharp images to small monitors on the desks. Dan Holbrook mounted two cameras in his yard tower so the yardmaster can see in both directions. Doug Tagsold used pairs of cameras mounted at key junctions so that a single operator could work several tower positions, **18**. And Ted Pamperin mounted a large computer screen behind a "window" by the dispatcher's desk so he can see out over the Chesapeake & Ohio's yard at Hinton, W.Va., just as his professional counterpart could do, **19**.

Passenger operations

After the wear and tear passenger equipment experienced during World War II, the railroads had a choice: assume the passenger train's golden era was over, or buy new passenger cars and locomotives.

That the heyday of passenger-train service was nearing an end wasn't that obvious in the late 1940s. The Interstate highway system wasn't on anyone's radar screen, the jet airliner had yet to make its debut, and automobiles closely resembled their pre-war cousins. Surely the traveling public would continue to use the safe, reliable, and smooth-riding passenger train as their primary intercity conveyance, especially after the new lightweight cars from Budd and Pullman-Standard arrived.

Modeling name trains and even many lesser trains is no longer a major challenge, as even a cursory perusal of the current Walthers catalog will reveal. The challenge is in using the passenger train in operationally interesting ways. If you have the space to model a major terminal, as Cliff Powers does, **7-1**, you can re-create the excitement as an endless parade of arriving and departing passenger trains. But even if you model a line that hosted but a handful of passenger trains per day, as I do, there are still ways to have the train do more than simply make a quick, uninterrupted tour of the railroad's main line.

It doesn't get much more basic than the daily trips of Nickel Plate Road St. Louis Division trains 9 and 10. On the Third Subdivision, which I model, these were nocturnal trains, passing through my hometown at 10:04 p.m. eastbound and 2:11 a.m. westbound.

Timetable 67, dated April 25, 1954, has a lower-case "s" to signify a scheduled stop for 9 and 10 for only three towns between the two division (and crew change) points. Since the two trains are first class and No. 10 is eastbound, it is superior to other trains, making for a rather uneventful run between terminals. And No. 9, inferior to 10 by direction, need be concerned only with its eastbound counterpart, and that meet normally occurs well east of the Third Sub.

But don't assume things went "by the book." I heard that Third Sub dispatchers routinely put 9 and 10 in the hole (passing track) for inferior freight trains, not to discourage ridership and hence support train-off applications to the Interstate Commerce Commission, but rather to avoid stopping and starting heavy freight trains. Trains 9 and 10 typically comprised one PA, a head-end car or two, a coach, diner, and a Pullman, so those trains were very easy to stop and start.

Adding to the wisdom of this approach is what might be called the Ten Mile Rule. Back in the days when a team of horses would pull a load of grain to the elevator for loading into a boxcar, elevators were spaced so that a farmer could get to the elevator, unload, and get back home again before dark during the fall harvest season. This meant that elevators were located about ten miles apart, which in turn meant

On my HO Nickel Plate layout, Frankfort "employs" a general yardmaster, two or three "footboard" yardmasters, a soybean plant job, a commercial engine, a roundhouse foreman, and an east-end staging yard crew. Add arriving and departing road crews, and the 4-foot-wide aisle gets crowded.

depots were located there, and probably passing tracks. So with a passing track every ten miles or so, a dispatcher could refine train meeting points to within a few minutes. Likely as not, the passenger train could enter a passing track at an interlocked switch, ease down the passing track as the freight roared by, and exit the other end of the pass without stopping via a spring switch. From a modeling point of view, what looked like a simple timetable meet became one that required train orders to be issued.

Throw in a few flag stops, easily orchestrated by printed message forms inserted into a train's instruction sleeve, and red home signals at one or more "dummy" interlocking plants along the route (thanks to Iowa Scaled Engineering's Interlocking in a Box), and an "easy run" becomes a distinct challenge to the passenger train's crew.

Although timetable and train-order operations were falling from favor as the transition era came to a close, many railroads still used this time-honored system for several decades thereafter. Centralized Traffic Control wasn't warranted on low-traffic-density lines, so until radios and work-rule changes made it feasible to issue track warrants, the timetable and train order reigned supreme.

Big city opportunities

If you model a major city, as Cliff Powers does for New Orleans and Chuck Hitchcock did with Kansas City on his former Santa Fe, there will be considerable head-end (mail, express, and baggage) work for the switching crew to perform. And even in smaller cities, passenger loading could cause a coach to be dropped or picked up. Depending on the time of day, a diner could also be dropped or picked up as a train neared the end of its run or was approaching the breakfast hour. With a little due diligence, you may find that a passenger train's operating potential can equal its physical allure.

Work trains

Trains that helped maintain the railroad were dispatched as "work extras." They comprised cars that held rail and other track supplies, carloads of ballast, **20**, even "camp cars" where the workers could stay overnight for jobs requiring more than a day.

Dispatching a work extra with orders specifying whether it will be protecting against extra trains, what times it can be expected to be working, and where it can add interest to an operating session. Just be sure the orders are issued to all trains the extra's work schedule and location(s) will affect.

One last caveat

The good news is that a realistically operating model railroad will draw a crowd of operators. The bad news is that a realistically operating model railroad will draw a crowd of operators, **21**—and, like relatives in the kitchen at Thanksgiving, that crowd will gather in one aisle. Plan ahead!

CHAPTER TEN

TRANSITION ERA PHOTO GALLERY

Jack Ozanich has skillfully captured all of the key ingredients of a transition era railroad, including adapting a classic EMD paint scheme to his fleet of covered wagons. The HO scale Atlantic Great Eastern is freelanced, but you'd never know it judging from its appearance or operations. *Craig Wilson*

Let's close our brief sojourn back into the last years of steam and the debut of colorful first-generation diesels with a selection of model and prototype photos that evoke that time and place.

For more information ... More helpful information and inspiring photographs of the transition era may be found in Kalmbach's *Classic Trains* magazine. CT's editors have also produced three reprints with expanded photo coverage of great articles about transition era railroading from back issues of *Trains* magazine titled *Trains of the 1940s*, *Trains of the 1950s*, and *More Trains of the 1950s.*

The paint schemes applied by the North Conway Scenic Railroad, above, provide a good example of adapting prototype designs to freelanced model railroads. The F unit has a Canadian National and Central Vermont look, while Geep 573 evokes the Boston & Maine's livery, handsomely modeled by E7 3810 on George Dutka's tribute to the B&M, left, shown existing the tunnel under downtown Bellows Falls, Vt. *Model photo: George Dutka*

Gerry Albers is modeling the Virginian Ry. in in the era when Fairbanks-Morse (idling in the shadows at right) had made its presence known in electrified territory with their brutish 2,400-hp Train Masters. The brass model of a Squarehead at left awaits a trip to the paint shop, but that can wait until the railroad's more pressing needs are met! *Gerry Albers*

The New York Central and Pennsylvania railroads interchanged in Sturgis, Mich. Above, grungy NYC H-6 Mikado 1890—a Trix model weathered using acrylic washes and powders by Jim Six—is about to drop off a cut of boxcars from Hillsdale, Mich. Note the extended coal-bunker sides. Pennsy H10 8304 is from Broadway Ltd. With coaching by Mikado expert Ray Breyer, Jim added an Elesco feedwater heater overhanging the smokebox front and Commonwealth trailing truck with booster to light Mikado 1853, top, the one NYC H-6 so equipped. It's also a Trix light 2-8-2. *Jim Six*

One of the most visually authentic tributes to the waning years of steam is on Warner Clark's Proto:48 (¼" fine scale) railroad. Above, former Lake Erie & Western, now Nickel Plate, Mikado 597 holds the main as a train led by a pair of inferior-by-direction Geeps eases into the siding at Malinta. This small Ohio town marks the spot where the NKP's Toledo Division crossed the Detroit, Toledo & Ironton. A DT&I SW1's crew, left, is picking up orders on the fly. *Richard Bourgerie*

Coal docks stand as reminders of another time. The photo at left, which dates to May 1971 when a Reading T-1 4-8-4 led a fan trip up the C&O's Greenbrier Branch to Durbin, W.Va., could easily have been taken two decades earlier. Another C&O coal dock at Peach Creek near Logan, W.Va., top, is almost identical to the HO Tichy kit, above, shown on my Allegheny Midland layout.

Thanks to hard-working individuals at museums across the continent, modelers have the chance to get up close and personal with many pieces of prototype equipment from the transition era. Pennsylvania GG1 4935, top, and EMD E7 5901, above—almost inconceivably, the only E7 that was saved from the scrapper's torch—now enjoy a sheltered existence in the superb Railroad Museum of Pennsylvania across the road from the Strasburg Rail Road. *Two photos: Doug Leffler*

Fan trips are another way to relive some aspects of the steam era. Chicago, Burlington & Quincy Mikado 4960 enjoyed an extended moment in the sun in excursion service in the 1960s. Its distinctive Q treatment with a high headlight and silvered graphite smokebox offer an example of adding a family look to an entire fleet of steam locomotives. It continues in service today on the Grand Canyon Ry, albeit with a much-altered appearance.

Beefy Santa Fe 2-10-2s 3841 and 1653 team up to shove an eastbound freight across Hwy. 138 at Pine Lodge, Calif., on Ted York's HO tribute to the AT&SF in 1952. Note the wig-wag signals protecting the crossing. *Ted York*

Chevrolet took the 1950s automobile tailfin craze one better by bending them over into arching wings for 1959. I took driver's training in a similar 1960 model, and that expanse of hood looked like an aircraft carrier from the driver's seat. You could load a small piano into the trunk.

RIGHT: Can't choose only a single prototype to model? More than one railroad was visible in many urban locations—here the Erie and the Delaware, Lackawanna & Western in Binghamton, N.Y. Harold Werthwein models the Erie's Delaware Division in the transition era, but this scene provided a good excuse to model some DL&W power.

Western Maryland's shops at Hagerstown, Md., were known as the "Hagerstown Hobby Shop" by railfans, as you could find everything from Alco FAs in both the speed-lettering and "circus" paint schemes, EMD BL2s, Alco RS-3s with dynamic brakes, a fleet of F units, and the occasional steam engine in excursion service. The FAs remind us that an easy way to suggest the passage of time is to have units in paint schemes from more than one era.

LACKAWANNA
861
LACKAWANNA
BUFFALO
RAILROAD
926

By the late 1950s, many "stock" diesels had been modified to fix wreck damage or improve performance. This Monon F3—note the three portholes—has an F7-style grille, making it an interesting detailing project. It retains its four "high-top" roof fans; later models had lower-profile fans.

Rhode Island's Narragansett Pier Ry. had an ornate two-story depot at Peace Dale and provided a home for an outdated wood clerestory-roof, open-platform passenger car. This is a good example of how one can model the transition era (or later) and still enjoy modeling cars dating back to the all-wood era.

Smaller power—here Rutland Consolidation no. 25 and Mikado no. 37 at Charlotte, Vt., on Randy Laframboise and Mike Sparks' HO railroad—teamed up to move heavy freights well into the 1950s. The lighter engine was almost always in the lead. Thanks to Digital Command Control, each engine can be assigned to its own engineer, or the two locomotives can be speed-matched and consisted. *Randy Laframboise*

We'll close our look back at the waning days of steam as the sun sets on an Allegheny Midland 2-6-6-2 easing into the yard at South Fork, W.Va., with today's harvest of "lake" coal from mines along the Otter Creek and Coal Fork subdivisions. At sunrise, it will be back at work in the verdant hills.